Chemical, Structural and Electronic Analysis of Heterogeneous Surfaces on Nanometer Scale

NATO ASI Series

Advanced Science Institutes Series

A Series presenting the results of activities sponsored by the NATO Science Committee, which aims at the dissemination of advanced scientific and technological knowledge, with a view to strengthening links between scientific communities.

The Series is published by an international board of publishers in conjunction with the NATO Scientific Affairs Division

A	**Life Sciences**	Plenum Publishing Corporation
B	**Physics**	London and New York
C	**Mathematical and Physical Sciences**	Kluwer Academic Publishers
D	**Behavioural and Social Sciences**	Dordrecht, Boston and London
E	**Applied Sciences**	
F	**Computer and Systems Sciences**	Springer-Verlag
G	**Ecological Sciences**	Berlin, Heidelberg, New York, London,
H	**Cell Biology**	Paris and Tokyo
I	**Global Environmental Change**	

PARTNERSHIP SUB-SERIES

1.	**Disarmament Technologies**	Kluwer Academic Publishers
2.	**Environment**	Springer-Verlag / Kluwer Academic Publishers
3.	**High Technology**	Kluwer Academic Publishers
4.	**Science and Technology Policy**	Kluwer Academic Publishers
5.	**Computer Networking**	Kluwer Academic Publishers

The Partnership Sub-Series incorporates activities undertaken in collaboration with NATO's Cooperation Partners, the countries of the CIS and Central and Eastern Europe, in Priority Areas of concern to those countries.

NATO-PCO-DATA BASE

The electronic index to the NATO ASI Series provides full bibliographical references (with keywords and/or abstracts) to more than 50000 contributions from international scientists published in all sections of the NATO ASI Series.
Access to the NATO-PCO-DATA BASE is possible in two ways:

– via online FILE 128 (NATO-PCO-DATA BASE) hosted by ESRIN,
Via Galileo Galilei, I-00044 Frascati, Italy.

– via CD-ROM "NATO-PCO-DATA BASE" with user-friendly retrieval software in English, French and German (© WTV GmbH and DATAWARE Technologies Inc. 1989).

The CD-ROM can be ordered through any member of the Board of Publishers or through NATO-PCO, Overijse, Belgium.

Series E: Applied Sciences - Vol. 333

Chemical, Structural and Electronic Analysis of Heterogeneous Surfaces on Nanometer Scale

edited by

Renzo Rosei

Sincrotrone Trieste, Italy
and Physics Department,
University of Trieste, Italy

Springer Science+Business Media, B.V.

Proceedings of the NATO Advanced Research Workshop on
Chemical, Structural and Electronic Analysis of Heterogeneous Surfaces
on Nanometer Scale
Trieste, Italy
24–26 April 1995

A C.I.P. Catalogue record for this book is available from the Library of Congress

ISBN 978-94-010-6414-9   ISBN 978-94-011-5724-7 (eBook)
DOI 10.1007/978-94-011-5724-7

Printed on acid-free paper

This book contains the proceedings of a NATO Advanced Research Workshop held within the programme of activities of the NATO Special Programme on *Nanoscale Science* as part of the activities of the NATO Science Committee.

Other books previously published as a result of the activities of the Special Programme are:

NASTASI, M., PARKING, D.M. and GLEITER, H. (eds.), *Mechanical Properties and Deformation Behavior of Materials Having Ultra-Fine Microstructures*. (E233) 1993 ISBN 0-7923-2195-2

VU THIEN BINH, GARCIA, N. and DRANSFELD, K. (eds.), *Nanosources and Manipulation of Atoms under High Fields and Temperatures: Applications*. (E235) 1993 ISBN 0-7923-2266-5

LEBURTON, J.-P., PASCUAL, J. and SOTOMAYOR TORRES, C. (eds.), *Phonons in Semiconductor Nanostructures*. (E236) 1993 ISBN 0-7923-2277-0

AVOURIS, P. (ed.), *Atomic and Nanometer-Scale Modification of Materials: Fundamentals and Applications*. (E239) 1993 ISBN 0-7923-2334-3

BLÖCHL, P. E., JOACHIM, C. and FISHER, A. J. (eds.), *Computations for the Nano-Scale*. (E240) 1993 ISBN 0-7923-2360-2

POHL, D. W. and COURJON, D. (eds.), *Near Field Optics*. (E242) 1993 ISBN 0-7923-2394-7

SALEMINK, H. W. M. and PASHLEY, M. D. (eds.), *Semiconductor Interfaces at the Sub-Nanometer Scale*. (E243) 1993 ISBN 0-7923-2397-1

BENSAHEL, D. C., CANHAM, L. T. and OSSICINI, S. (eds.), *Optical Properties of Low Dimensional Silicon Structures*. (E244) 1993 ISBN 0-7923-2446-3

HERNANDO, A. (ed.), *Nanomagnetism* (E247) 1993. ISBN 0-7923-2485-4

LOCKWOOD, D.J. and PINCZUK, A. (eds.), *Optical Phenomena in Semiconductor Structures of Reduced Dimensions* (E248) 1993. ISBN 0-7923-2512-5

GENTILI, M., GIOVANNELLA, C. and SELCI, S. (eds.), *Nanolithography: A Borderland Between STM, EB, IB, and X-Ray Lithographies* (E264) 1994. ISBN 0-7923-2794-2

GÜNTHERODT, H.-J., ANSELMETTI, D. and MEYER, E. (eds.), *Forces in Scanning Probe Methods* (E286) 1995. ISBN 0-7923-3406-X

GEWIRTH, A.A. and SIEGENTHALER, H. (eds.), *Nanoscale Probes of the Solid/Liquid Interface* (E288) 1995. ISBN 0-7923-3454-X

CERDEIRA, H.A., KRAMER, B. and SCHÖN, G. (eds.), *Quantum Dynamics of Submicron Structures* (E291) 1995. ISBN 0-7923-3469-8

WELLAND, M.E. and GIMZEWSKI, J.K. (eds.), *Ultimate Limits of Fabrication and Measurement* (E292) 1995. ISBN 0-7923-3504-X

EBERL, K., PETROFF, P.M. and DEMEESTER, P. (eds.), *Low Dimensional Structures Prepared by Epitaxial Growth or Regrowth on Patterned Substrates* (E298) 1995. ISBN 0-7923-3679-8

MARTI, O. and MÖLLER, R. (eds.), *Photons and Local Probes* (E300) 1995. ISBN 0-7923-3709-3

GUNTHER, L. and BARBARA, B. (eds.), *Quantum Tunneling of Magnetization - QTM '94* (E301) 1995. ISBN 0-7923-3775-1

PERSSON, B.N.J. and TOSATTI, E. (eds.), *Physics of Sliding Friction* (E311) 1996. ISBN 0-7923-3935-5

MARTIN, T.P. (ed.), *Large Clusters of Atoms and Molecules* (E313) 1996. ISBN 0-7923-3937-1

DUCLOY, M. and BLOCH, D. (eds.), *Quantum Optics of Confined Systems* (E314). 1996. ISBN 0-7923-3974-6

ANDREONI, W. (ed.), *The Chemical Physics of Fullereness 10 (and 5) Years Later. The Far-Reaching Impact of the Discovery of* C_{60} (E316). 1996. ISBN 0-7923-4000-0

NIETO-VESPERINAS, M. and GARCIA, N. (Eds.): *Optics at the Nanometer Scale: Imaging and Storing with Photonic Near Fields* (E319). 1996. ISBN 0-7923-4020-5

RARITY, J. and WEISBUCH, C. (Eds.): *Microcavities and Photonic Bandgaps: Physics and Applications* (E324). 1996. ISBN 0-7923-4170-8

LURYI, S., XU, J. and ZASLAVSKY, A. (Eds.): *Future Trends in Microelectronics: Reflections on the Road to Nanotechnology* (E323). 1996. ISBN 0-7923-4169-4

JAUHO, A. and BUZANEVA, E.V. (Eds.): *Frontiers in Nanoscale Science for Micron/Submicron Devices* (E328). 1996. ISBN 0-7923-4301-8

Table of Contents

Preface by Prof. Renzo Rosei, Director of NATO ARW

The NATO Advanced Research Workshop "Chemical, structural and electronic analysis of heterogeneous surfaces on nanometer scale" was held in Trieste on 24-26 April 1995. It was aimed at identifying recent achievements and relative strengths of two developing techniques for characterisation of surface properties on the nanometer scale, namely: (i) local probe methods (LPMs), including scanning tunnelling microscopy (STM) and its derivatives; and (ii) nano-scale photoemission and absorption spectroscopy for chemical analysis.

The programme consisted of ten key lectures delivered by scientists who are at the forefront of developing and applying of the two forms of microscopy, and of six oral presentations.

In these proceedings Prof. Ph. Avouris (IBM, Thomas J. Watson Research Center, USA) deals with the possible application of STM as a tool for atomically resolved chemical analysis and describes in particular the imaging of defects or atomic adsorbates using STM in its spectroscopy mode to probe electron scattering and desorption of individual atoms as a result of excitations induced by the tunnelling current.

The paper of Prof. O. Marti (University of Ulm, Germany) describes the principles of scanning force/friction and scanning near-field optical microscopes (SFFM and SNOM) and their application for characterisation of heterogeneous materials with domain sizes below 1 mμ.

The current status and the recent results of the scanning photoemission electron microscopes built at ELETTRA (Trieste) and SRRC (Taiwan) are presented in the paper of Prof. G. Margaritondo (Ecole Polythèchnique Fèdèrale, Lausanne, Switzerland & Sincrotrone Trieste), who reviews critically the main features of synchrotron radiation spectromicroscopy analysing the strategies for optimisation and the milestones in the future development of this powerful analytical method.

The paper of Prof. H. Ade (University of North Carolina, U.S.A.) reviews the recent progress in development of spatially-resolved photoelectron spectroscopies and in particular, is focused on the strategies employed in construction of scanning microscopes using zone plate photon optics, their advantages and disadvantages and their applications in material science and biology using both photoemission and absorption modes.

Prof. E. Bauer (University of Clausthal, Germany) gives an excellent review of the present state of the non-scanning photoemission microscopy with slow electrons which has the attractive advantage of combining comprehensive structural with chemical analysis on the 0.01-0.1 mμ scale.

The paper by Dr. R. Fink (University of Stuttgart, Germany) describes an overview of the project presented at BESSY 2 for a non-scanning photoelectron microscope where using electron optics spatial resolution of 0.01 mμ will be possible.

Dr. H. Rotermun (FHI Berlin, Germany) presents the latest (spectacular!) results of spatially resolved ($\leq$ 1 mμ) in-situ reaction studies of chemical waves and oscillatory phenomena obtained with the UV-photoemission electron microscope and gives the latest development of this microscope using optical methods like ellipsometry.

The Workshop was a strong success in bringing two communities of microscopists closer together and in emphasizing the opportunities that the two major techniques which have been presented will provide, giving complementary information on a wide range of nano-science phenomena. It is my hope that the present Proceedings will convey in the reader the same enthusiasm and the same sense of opportunities which was high and alive in Trieste.

SURFACE STATE ELECTRONS: TRANSPORT THROUGH DANGLING BONDS ON SILICON, AND SCATTERING AND CONFINEMENT ON METALS

PH. AVOURIS, I.-W. LYO, and Y. HASEGAWA*
IBM Research Division, T.J. Watson Research Center
P.O. Box 218, Yorktown Heights, New York 10598, U.S.A

1. Introduction

Scanning tunneling microscopy (STM) and spectroscopy (STS) [1] continue to find new applications. They are currently used to probe at the atomic scale, metals, semiconductors, superconductors, molecular and biological systems, and adsorbed layers [2]. Here we present two new applications in the study of two- and lower-dimensional electron systems involving electrons in surface states of semiconductors and metals. First, STM tip-sample electrical point-contacts are used to probe for electron transport among the dangling-bond states of Si(111)-7x7, and study the electrical properties of model nanostructures. Then, we show how one can take advantage of the wave character of the electron and use electron interference effects to study the interaction of the quasi-two-dimensional free-electron gas (2DFEG) formed by Shockley metal surface states with individual surface steps and adsorbed atoms. We use spatially-resolved surface state spectroscopy to probe: the electronic structure of the steps, bulk-surface state mixing near them, the nature of the images of adsorbed atoms, and their interaction with surface states. Finally, we use the fact that surface steps act as barriers for electrons to confine them, form lower-dimensionality structures, and observe quantum size effects at room temperature.

The Si(111)-7x7 surface has long been a puzzle. Its complex atomic structure has been the subject of numerous investigations and was finally determined by electron diffraction experiments [3]. However, STM was also crucial in the elucidation of its atomic structure [1]. The electronic structure of Si(111)-7x7 is also intriguing, and has attracted significant attention. Photoemission [4] and inverse-photoemission [5] spectroscopies indicate that the surface, so-called Si adatom-layer, has a finite density of occupied and unoccupied states at E_F, and STS-spectra [6] corroborate this conclusion. However, the dangling-bonds of the adatom-layer, are spaced far apart, (7 Å), and no dispersion is observed by angle-resolved photoemission [4], a fact that indicates a very weak interaction between the dangling-bonds. Attempts to directly measure

1

R. Rosei (ed.),
Chemical, Structural and Electronic Analysis of Heterogeneous Surfaces on Nanometer Scale, 1–23.
© 1997 IBM.

electrical transport involving these states have been unsuccessful [7]. However, the anomalous broadening of the quasi-elastic peak observed in electron energy loss studies of Si(111)-7x7 has been interpreted as indicating the presence of surface ac-conductivity [8].

Here we report on our efforts to directly measure electrical transport through the dangling-bond states of 7x7. In our approach to this problem, we decrease the importance of bulk transport by using rectifying, not ohmic, metal-semiconductor contacts, and by increasing the spreading resistance (ρ ~1/(radius of contact area)) through the use of the ultra-sharp STM-tip. Surface and bulk contributions to the electrical transport are then resolved by exposing the sample to gases which are known to chemically react with the surface dangling-bonds and thus affect surface conduction. In this way, we are able to observe conduction involving dangling-bond surface states.

We also use STM tip-sample contacts to electrically address individual nanostructures. While there is considerable interest in the electrical properties and quantum size effects occurring in nanostructures, there has not been a way to establish proper electrical contacts with such nanostructures. The ability of the STM to image a set of nanostructures, select a particular one and electrically address it through a point-contact make the STM an ideal tool for such studies.

Electrons in surface states of metals behave, in general, very differently from electrons in dangling-bond states of semiconductors. Of particular interest are s,p-electrons in the so-called Shockley surfaces states. In these states, the confinement of the electrons to the surface region is brought about by a projected bulk band-gap on one side and by the vacuum barrier on the other side. The confinement is not strictly to the top layer but there is some bulk penetration. Most importantly, however, the electrons are free to move parallel to the surface, forming a quasi-two-dimensional free-electron gas (2DFEG). Typical cases of Shockley surface states are provided by the late transition and noble metals Ni, Pd, Pt, Cu, Ag, Au, where band gaps exist in the sp bands at the L and X points of the bulk Brillouin zone [9]. There is renewed interest in these surface states because of evidence that they affect a great variety of surface properties and processes, such as the range of adsorbate-adsorbate, adsorbate-step, and step-step interactions, the sticking coefficient of adsorbates, the magnitude of activation barriers for dissociative chemisorption, damping of adsorbate vibrations, second harmonic generation, etc..

Because of the particle-wave duality, an electron, acting as a wave, can interfere with itself. A surface state electron incident on one of the above obstacles is partially reflected. The reflected and the incident wave then interfere to give an oscillatory local density-of-states (LDOS (E,r)) in the vicinity of the obstacle. The first attempt to study such electron interference effects with STM was made by Jaklevic and co-workers [10]. More recently, two IBM groups [11,12] were able to image directly the LDOS oscillations with STM and STS. Here we present results illustrating the reflection and interference induced by steps and adsorbed atoms at Au(111) and Ag(111) surfaces. We investigate the spatial extent of the electron-step and electron-adsorbate interactions by studying the surface state spectrum as a function of the distance from steps and adsorbates. Since surface steps act as barriers for surface electrons, we have used narrow terraces and metal islands to confine these electrons and form low dimensionality structures. With the STM we are able to image the probability distributions of the particle-in-a-box-like states formed and obtain their discrete spectra. Because of the ultra-small size of these structures, we can observe quantum size effects even at room temperature.

2. Experimental

All experiments were carried out at room temperature in ultrahigh vacuum (base pressure $<2 \times 10^{-10}$ Torr). The Si(111) and Si(100) samples were commercial, polished wafers (1 Ω cm P-doped and 0.1 Ω cm B-doped). The native oxide was removed either by thermal flashing or by HF etching followed by hydrogen thermal desorption. Si epitaxial islands were grown by evaporating silicon on a Si(111) substrate held at 400∞C. The Au(111) samples were prepared in situ by evaporating gold on cleaved mica at room temperature, and subsequent annealing at 600∞C. W tips were used for imaging, while both W and Al tips were used in point-contact spectroscopy (I-V measurements).

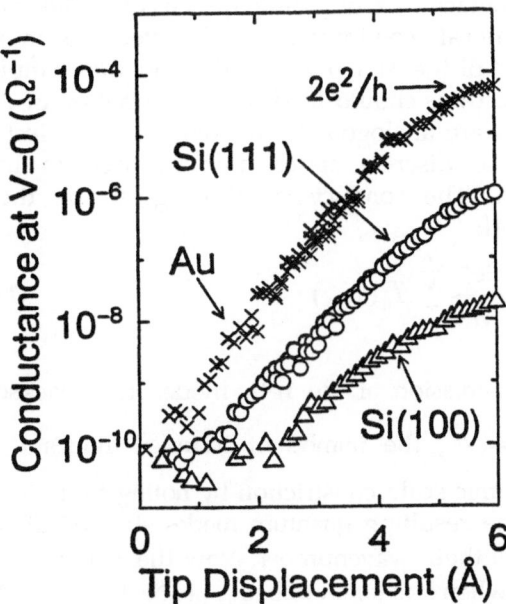

Fig. 1. Conductance at V=0 plotted as a function of the displacement of the tip from the tunneling position. Measurements are shown for three samples: polycrystalline Au(111), a Si(111)-7x7 surface, and a Si(100)-2x1 surface.

4

The STM and the scanning tunneling microscopy (STS) setups have been described elsewhere [13]. Control of the STM tip motion is accomplished by adding D/A-converter voltages to the z-piezo voltage. I-V measurements as a function of tip-surface distance are performed by disabling the feedback loop, displacing the tip, typically by 0.1 Å, and ramping the bias voltage; then the sequence is repeated. I-V data are continuously collected during both the approach and retraction of the tip from the surface. In this way, we can monitor any changes in the electronic structure of the tip or sample.

3. Results and Discussion

3.1. THE STM AS A PROBE OF ELECTRICAL TRANSPORT

3.1.1. *Electrical Properties of Nanometer-Scale Contacts*
In Fig. 1, the differential conductance (dI/dV) at V=0 is plotted as a function of tip-sample distance for three different materials: a polycrystalline gold sample, and single crystal Si(111)-7x7 and Si(100)-2x1 samples. Zero tip displacement corresponds to the tip position during normal tunneling operation, while a tip displacement of about 6 Å leads to tip-sample contact. As Fig. 1 shows, when first contact is established with the gold sample, the value of the conductance is very close to $2e^2/h$. When the tip is further pressed against the sample, the value of the conductance increases. A conductance of $2e^2/h$ is characteristic of conduction through a single ideal transport channel [14], and suggests a contact area of atomic dimensions. This can be understood by considering the lateral confinement of the electrons in the constriction formed by the contact of the sharp metal tip and the sample. This lateral confinement of the electrons should lead to the formation of discrete quantum levels (modes) which are analogous to the modes of an electron waveguide. Electrons occupy these discrete states and are free to move along the constriction direction. The conductance G is given by the two-terminal Landauer equation [14]:

$$G = \frac{2e^2}{h} \sum_{n=1}^{N} T_n(E_F) \tag{1}$$

where T_n is the transmission of the n-th mode. If no backscattering occurs, then: $\sum_{n=1}^{N} T_n(E_F) = N$, the number of occupied modes. We can estimate this number for an atomic scale constriction by noting that, if W is the width of the constriction, the resulting quantum modes should satisfy the relation $W=n(\pi/k)$, where k is their wavenumber. Now the number of occupied modes N will be given by n when k is taken to be k_F, i.e. $N=Int(\frac{k_F}{\pi})$. So for typical values of W=2.5 Å, k_F =1.5 Å^{-1}, N is 1 and the expected conductance will be $2e^2/h$ as found here. Although, the above discussion is somewhat simplistic, it does give credence to the assertion that atomic-scale contacts can be produced with the STM.

While the electrical properties of the tip contact to the metal sample can be understood in terms of constriction resistance, the properties of the contacts to the two silicon surfaces behave quite differently. The conductance values are two-orders and four-orders of magnitude lower than that of gold for Si(111)-(7x7) and Si(100)-(2x1), respectively. The contact conductance of Si(111)-(7x7) is different from that of Si(111)-(2x1) even when the two samples have the same bulk conductivity. Moreover, the conductance is independent of the doping, i.e. both p- and n-doped samples have the same contact conductance. These observations suggest the involvement of a surface transport channel. This notion is further supported by the fact that the observed conductances are much larger than the value expected (less than 10^{-12} ohm^{-1}) if transport was to be limited by the Schottky barrier that should form at the tip (tungsten)-silicon interface [15].

3.1.2. *Surface Electrical Transport on Si(111)-(7x7)*

The involvement of the surface in the electrical transport was confirmed in an experiment in which, after the electrical contact was formed, the Si(111) surface was exposed to gases that react with the surface dangling-bond states and the change in conductance was measured [16]. Fig. 2 shows the behavior of a W-Si(111)-(7x7) contact after exposure to O_2. It is seen that after an exposure of ~5L of O_2, the conductance decreased by about three orders of magnitude. The same reduction was observed when an aluminum tip was used.

More information about the nature of the point contacts was obtained from an analysis of their I-V characteristics. In Fig. 3 we show the characteristics of an ~5nm diameter W-Si(111)-(7x7) contact (the diameter was determined by breaking the adhesive contact and imaging the affected area of the surface). The I-V's show strong rectification behavior with the contact acting as a nanometer size Schottky-diode.

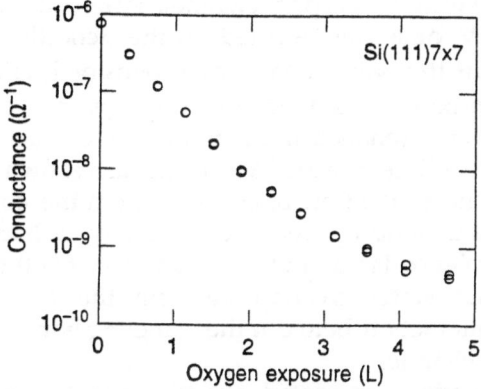

Fig. 2. Dependence of the conductance measured with the tip in contact with a Si(111)-7x7 surface as a function of exposure to O_2.

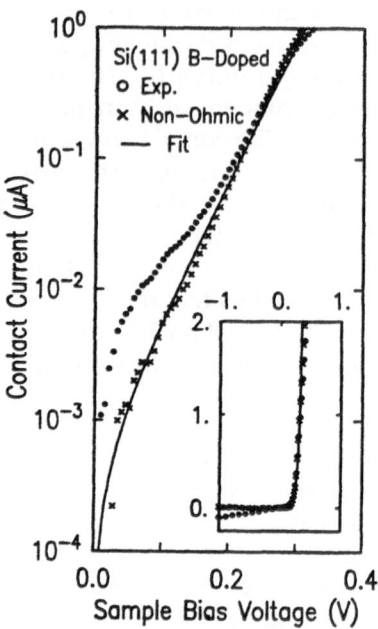

Fig. 3. Current vs. sample bias voltage of a point-contact between a W tip and a Si(111)-7x7 (B-doped) sample. The experimental data are shown as open circles, while filled squares give the current after the subtraction of the ohmic component. The solid line gives the fit to the Schottky equation. Inset: I-V curves of the point-contact.

For reverse bias we observe an ohmic "leakage" current (Fig. 3). We find that it is this linear component which is most sensitive to O_2 exposure, and, thus, can be attributed to the surface transport channel. After subtraction of this "leakage" current, the I-V data can be fitted to the Schottky equation $J=J_s$ (exp$\propto$ V/kT>-1) to obtain the saturation current density J_s and from it the Schottky barrier, ϕ_{SB}, can be extracted. In this way, we can determine the relation of the valence and conduction bands at the contact area to the corresponding bands in the adjacent area. The results are shown in Fig. 4. We see that barriers are present for the flow of electrons from the conduction band of the contact area to the surrounding area, and, similarly, a barrier exists for holes to flow out of the valence band of the contact area. On the other hand, the dangling-bond surface states overlap and pin the Fermi level and, provided that there is interaction between the dangling-bonds, current could easily flow through this channel.

The experimental observations presented above prove the existence of a surface transport channel, and, moreover, allow the identification of the nature of this channel. In principle, there are two possibilities: One involves transport through the dangling-bond surface state, while the other involves the space-charge layer present due to surface band-bending.

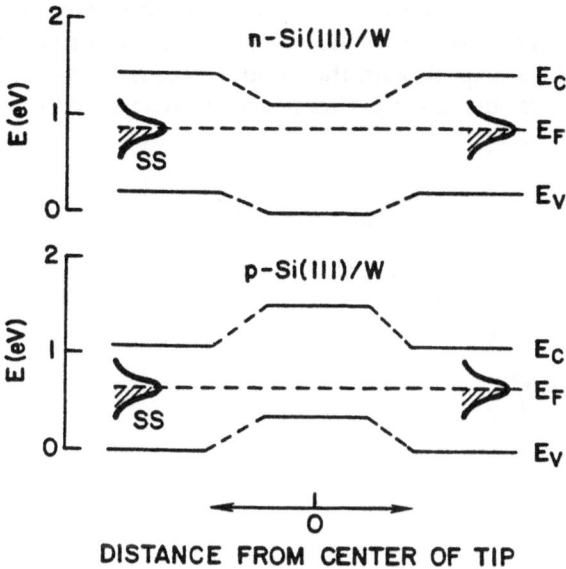

Fig. 4. Diagrams showing the energies of the top of the valence band and the bottom of the conduction band of an n- and a p-doped sample in contact with a W tip. The area of the surface in contact with the W tip is depicted in the middle of the diagram. The surface states (dangling-bonds) in the surrounding clean 7x7 surface pin the Fermi level.

To our knowledge, previous work on semiconductor surface transport has always implicated transport involving the space-charge layer. The evidence presented here, however, is consistent with transport via the surface state. Specifically, the finding that the contact resistance is independent of the bulk doping cannot be reconciled with a transport mechanism involving the space-charge layer. Moreover, the very large, by a factor of ~10^3, reduction in conductance upon O_2- exposure, would require a very strongly accumulated or inverted space-charge layer to be present at the surface of Si(111)-7x7. However, it is well established that the Fermi level is pinned by the surface states at 0.6eV about the valence band maximum [17], leading to an electron depletion layer for n-type 7x7, and a hole depletion layer for p-type 7x7. Finally, as we discussed above, the electronic structure shown in Fig. 4 provides another reason why transport through the surface state should be important. More evidence will be provided when we next consider the behavior of tip point-contacts to model nanostructures.

As a simple model of a nanostructure whose electrical properties we want to determine by tip-sample point-contacts we have selected small Si epitaxial bilayer islands grown on Si(111). These islands are formed in various sizes when Si is deposited on Si(111). The advantage of these epitaxial structures is that the coupling of the bilayer to the bulk is the same as that of the bilayer of the substrate. The experiment then consists of measuring and comparing the conductances obtained by tip point-contact on top of an island and on the adjacent flat substrate. The two conductances measured in this way are found to be different. Although, the exact numerical value of the difference varies somewhat form experiment to experiment, the conductance over the island is always smaller. The results of such an experiment are shown in Fig. 5.

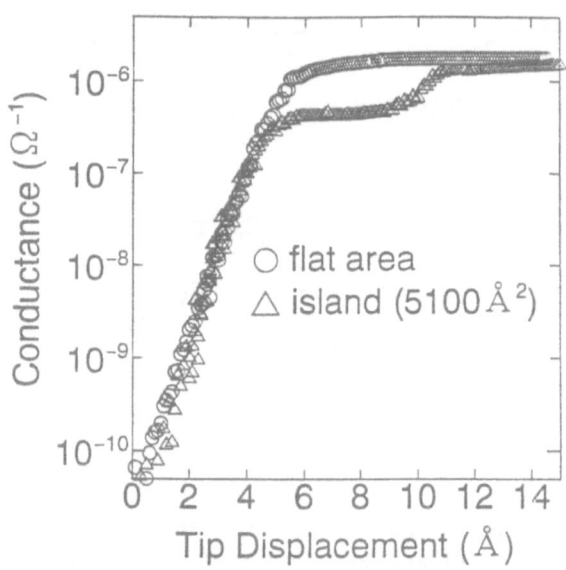

Fig. 5. Conductance at V=0 plotted as a function of displacement of the STM tip from the tunneling position towards contact, measured on an epitaxial Si(111) island and on the surrounding Si(111) terrace.

It is seen that the conductance over the island measured by the initial tip contact is smaller, but by continuing to press the tip against the sample with the z-piezo transducer, the conductance increases and reaches that of the flat substrate surface.

Given that coupling to the bulk is the same for both the island and substrate, the above findings can be explained by invoking transport through the dangling-bond surface state. One expects that the steps surrounding the island would act as barriers to such transport scattering the carriers and leading to a reduced conductance [16]. By pressing the tip against the island, however, the contact area increases and at some point contact with the surrounding substrate is achieved and the conductance becomes the same as for direct contact to the substrate.

While the carrier scattering processes at the boundaries of the island are studied indirectly from their effect on the value of the contact conductance, electron scattering processes at metal steps can be studied more directly by observing electron interference patterns.

3.2. ELECTRON SCATTERING AND CONFINEMENT ON METAL SURFACES

3.2.1. *Scattering Steps and Adsorbates*

Electrons in Shockley surface states of metals provide a new quasi-two-dimensional electron gas (2DFEG) system of higher electron density than those provided by inversion layers in semiconductors [18] or semiconductor heterostructure interfaces [19]. It is not, however, obvious how to study such a system with STM/STS. A true 2D electron system has a constant density-of-states (DOS) so, at first glance, STS should not be of much help. However, what is of most interest here is the interaction of the 2DFEG with scattering potentials produced by adsorbed or embedded atoms or structural features such as surface steps [20]. As we will see below, these types of interactions can be studied by STS.

The Shockley surface states that we study here involve the s,p-electrons of the (111) surfaces of Ag and Au. In Fig. 6 we show STS spectra of a Au(111) surface near the onset (E_0) of its surface state. The full line gives the spectrum obtained over a terrace site of the Au(111) surface and clearly shows the onset of the surface state at about 0.4 eV below E_F. When the tip is placed directly over a surface step, however, the characteristic onset of the surface state is absent. We observe an analogous behavior near point-defects and adsorbates.

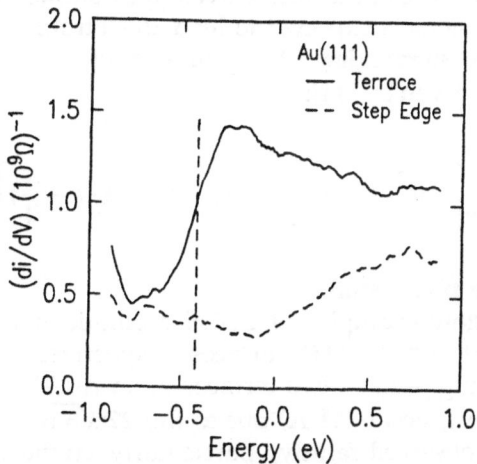

Fig. 6. Tunneling spectra obtained over a terrace site of a Au(111) surface (full line) and directly over a step of the same surface (dotted line).

The fact that the surface state is not observed in the vicinity of the above surface features can be phenomenologically described by considering the step or adsorbate to act as a electron barrier. The picture is analogous to that used to model electrical resistivity due to point and extended defects [21]. The de Broglie wavelength of the electrons at the noble metal surfaces is significantly larger than that in the bulk of the materials. At the Fermi energy the surface electron wavelengths for Cu(111), Ag(111) and Au(111) are 29 Å, 49 Å and 36 Å, respectively [9]. When an electron wave is incident on a step or defect, it will be scattered. The reflected part of the wave can then interfere with the incident part, resulting in an oscillatory local density-of-states (LDOS), i.e. a standing wave in the vicinity of the step [11,12]. In the case of two-dimensional scattering, the resulting LDOS oscillations should, in the absence of other dephasing processes, decay slowly as a function of the distance, x, away from the step. If we assume that the scattering potential can be modeled by a hard-wall, then the resulting local-density-of-states $\rho(E,x)$ is given by:

$$\rho(E,x) \propto 1 - J_0(2k_{//}x) \qquad (2)$$

where J_0 is the zeroth-order Bessel function, and $k_{//}$ is the wavenumber parallel to the surface, i.e. $k_{//} = (k_x^2 + k_y^2)^{1/2}$. The decay of the oscillations with distance x from the step can be seen from the asymptotic form of the Bessel function:

$$J_0(2k_{//}x) \cong (\frac{1}{\pi k_{//}x})^{1/2} \cos(2k_{//}x - \frac{1}{4}\pi) \qquad (3)$$

In contrast to steps, scattering of the long wavelength surface-waves by point-defects of atomic dimensions is expected to lead to isotropic scattering giving rise to circular waves surrounding the point-defect. Within the s-wave approximation, r(E,r) is given by [11]:

$$\rho(E,) \propto 1 + \frac{2}{\pi k_{//}r}\left[\cos^2(k_{//}r - \frac{\pi}{4} + \eta_0) - \cos^2(k_{//}r - \frac{\pi}{4})\right] \qquad (4)$$

where h_0 is the s-wave phase-shift.

In Figs. 7 and 8 we show examples of LDOS oscillations observed at room temperature on Au(111) and Ag(111) surfaces, respectively. Fig. 7(a) shows the STM image containing parts of two terraces of an Au(111) surface (sample bias -1V). A small corrugation (~0.1 Å) due to the 22x√3 reconstruction on the Au(111) surface [22] is observed faintly, particularly on the lower terrace. In addition to the topographic image, we have obtained tunneling spectra at 3 Å intervals over the entire area shown in the image of Fig. 7(a). Using these I-V data, spatial images of the derivative dI/dV or (dI/dV)/(I/V) at given voltages, referred to as spectroscopic maps, can be produced. Such a map for V=+0.15 V is shown in Fig. 7(b). Here, bright contrast indicates a large value of the current derivative. Since the value of dI/dV provides a rough measure of LDOS, we can consider the image of Fig. 7(b) as a map of the surface LDOS at an energy of about 0.15 eV above E_F. This image shows a bright line, i.e. high LDOS, right at the step, and behind the step on the upper terrace side several more bright lines running parallel to the step are seen.

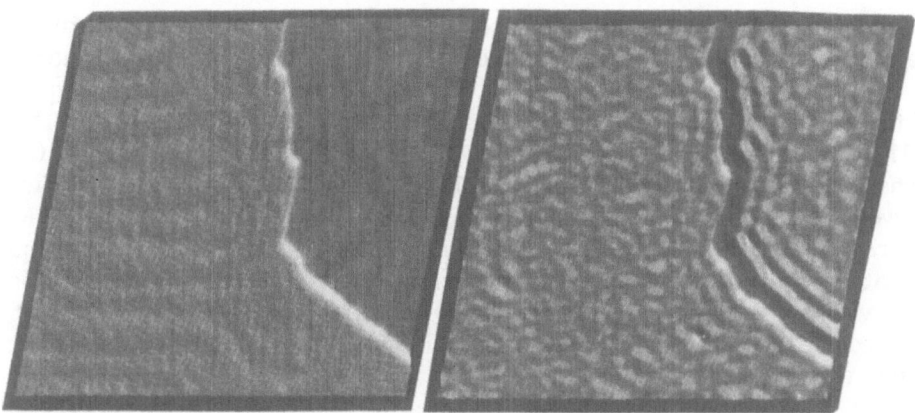

Fig. 7. Top: STM image of a Au(111) surface showing two terraces (sample bias=-1V). Bottom: Spectroscopic image at V_s =+0.15V, constructed from I-V curves obtained at 3 Å intervals over the same area shown in the STM image.

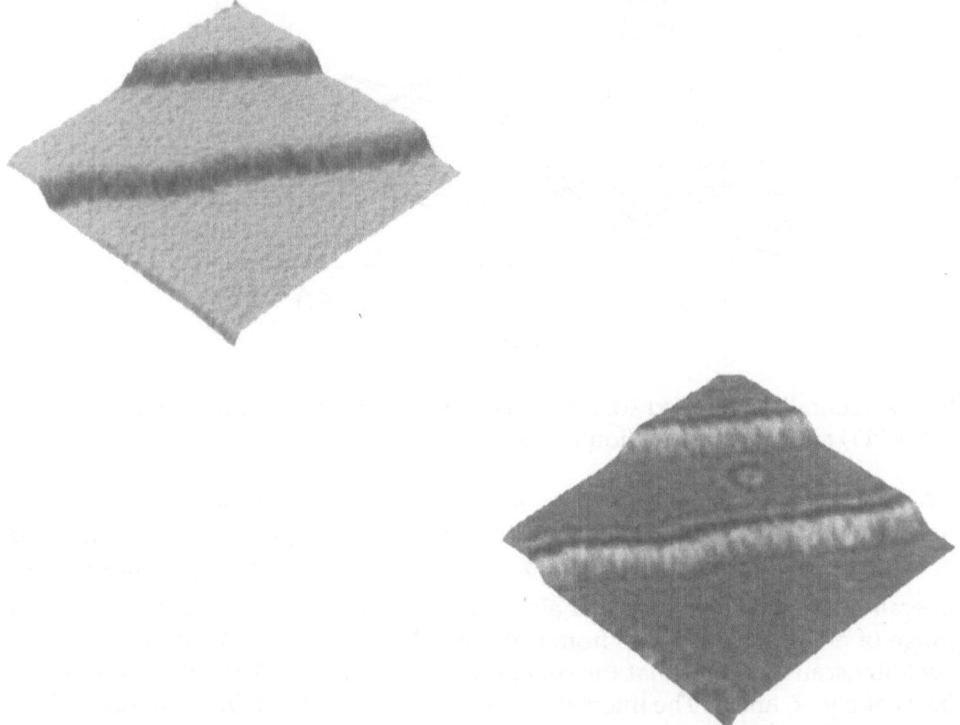

Fig. 8. Top: STM image of a Ag(111) surface showing areas of three terraces (V_s =+0.17V). Bottom: Spectroscopic image of the same area superimposed on the STM image (V_s =+0.44V). The arrow points to an impurity atom.

The spacing between the lines is about 16 Å. These observations are in agreement with an electron interference origin of the observed oscillations. The distance between two antinodes of the interference pattern should equal half the electron wavelength which, according to the known dispersion of the Au(111) surface state [9], should be about 31 Å at 0.15eV above E_F. In Fig. 8(a) we show an STM topograph of an Ag(111) surface at +0.17V. While, in Fig. 8(b) we give the corresponding spectroscopic map (superimposed on the STM topograph) obtained at +0.44 V. This map shows not only the oscillations associated with the steps, but also scattering by an impurity atom which is barely visible as a dark spot (see arrow) in the topographic image of Fig. 8(a). As expected, the point-defect produces oscillations in the form of concentric circular waves.

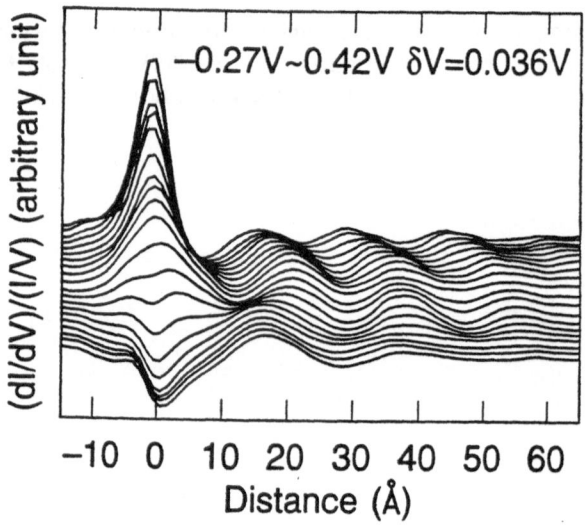

Fig. 9. Laterally-averaged (dI/dV)/(I/V) linescans perpendicular to a step on a Au(111) surface as a function of the bias.

Another way of displaying the standing waves involves STM line-scans normal to the direction of the step. To improve the signal-to-noise ratio, and account for step atom motion at room temperature, we average these scans laterally. Fig. 9 shows linescans along the $<1\,\overline{12}>$ direction of Au(111) for a range of energies (voltages) from 0.38 eV below E_F to 0.45 eV above E_F. From such linescans we find that the oscillations decay faster than expected on the basis of eqs. 2, and 3. The intensity decays exponentially with distance x from the step: $I \propto \exp(-x/\ell)$. This behavior was explained [11] as reflecting the fact that STM does not probe a single k-level, but rather an electron wave-packet with a spread determined by the temperature-dependent energy resolution ΔE. For a parabolic band $\Delta k_{//} = (m^* \Delta E)/(\hbar^2 k_{//})$, with

$\Delta E \simeq 3.5 k_B T$. The coherence-length ℓ_c is $\ell_c \simeq 1/\Delta k_{//}$. The observed $\ell_c \simeq 30$ Å is in rough agreement with these expectations [11]. From dI/dV line-scans as a function of bias voltage one can also determine the surface state dispersion near $k_{//} = 0$.

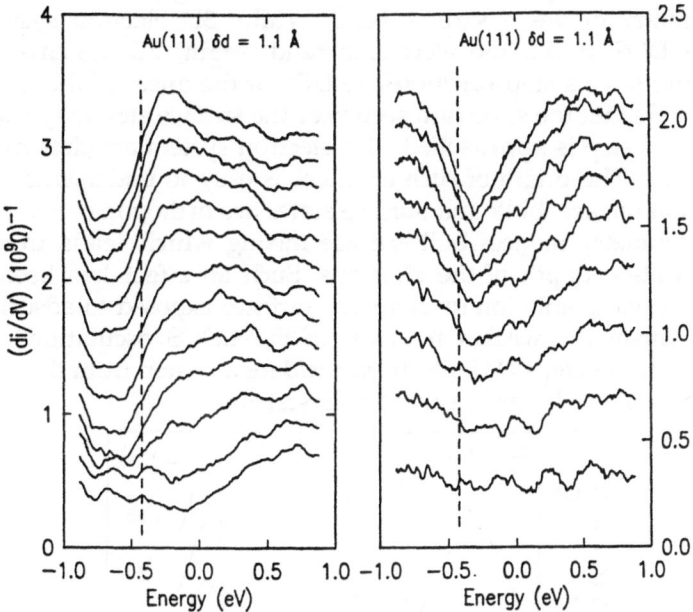

Fig. 10. (A) Tunneling spectra of the onset region of the surface state of Au(111) as a function of the distance from a step. The bottom spectrum is taken directly over the step while the rest are taken at 1.1 Å intervals. (B) Difference of the spectra obtained at distances x>4 Å from the step, from that obtained at x=4 Å.

To obtain more information on the nature and the spatial extent of the interaction of surface state electrons with steps, we have performed a different type of experiment where we monitor differences in surface-state spectra measured as a function of the distance from the step edge. Since surface state spectra are featureless (a true two-dimensional system has a constant DOS at energies above the onset energy [18]), we focused our attention on the onset region. In Fig. 10(A) we show a number of such spectra as a function of the distance, on the upper terrace side, from the step-edge of a Au(111) surface. The bottom curve corresponds to the spectrum obtained directly over the step edge, and the distance between successive measurements is 1.1 Å. From the changes in the shape of the spectra shown in Fig. 10(A) we conclude that the perturbation on the onset region of the spectra produced by the step extends at least 10 Å away from it. Moreover, the nature of the spectral modification is characteristic. In Fig. 10(B) we show the difference of the spectra obtained at distances larger than 4 Å from the step, from that obtained at x=4 Å. The distance of 4 Å was chosen to be close but not exactly at the step edge where

14

there is no evidence of a surface state. From these different spectra we see that the main effect of the step is to reduce the LDOS near the onset region (dashed line) of the surface-state band and to produce a more gradual increase of the LDOS with increasing electron energy. The onset region is most sensitive to the dimensionality of the electronic band. While 3D electron systems have a vanishing DOS at the the electronic band origin, 2D systems such as the surface state have a step-function-like DOS at the onset of the band [18]. It is thus plausible that the spectrum seen near the surface step may indicate that, as the surface step is approached, the electron system acquires increasingly a 3D character. The origin of such an effect is easy to understand. While, on a perfectly flat terrace, bulk and surface states are orthogonal, near a defect the reduced symmetry induces bulk-surface mixing which leads to an increased bulk penetration depth of the electrons. Such an effect has been invoked in interpreting photoemission results from surfaces exposed to adsorbates [23]. It can also explain the weaker intensity of the LDOS oscillations observed on the lower sides of steps [11]. Electrons incident at a step from the lower terrace side can be scattered more easily to bulk states.

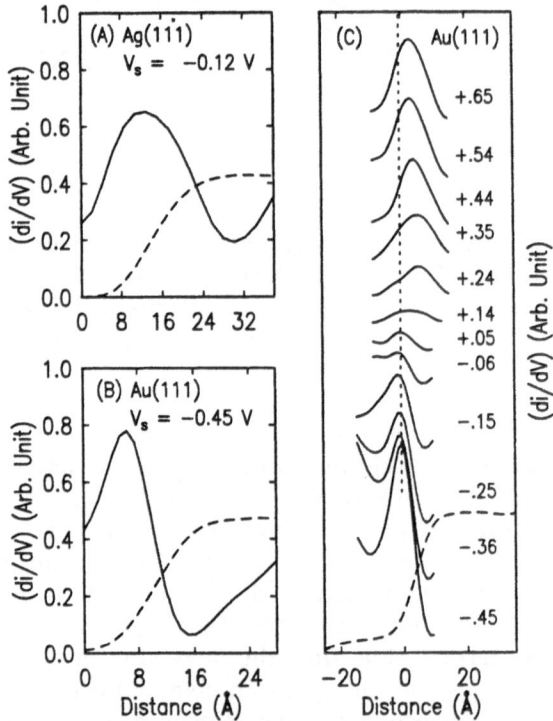

Fig. 11. (A) Solid line: dI/dV line-scan across a monoatomic step on a Ag(111) surface. Dashed line: constant current STM line-scan across the same step. The sample bias (Vs) is -0.12V. (B) Solid line: dI/dV line-scan across a step on a Au(111) surface. Dashed line: STM line-scan across the same step (V_s =- 0.45V). (C) Shift of the position of the dI/dV peak (step-edge state) at a Au(111) surface as a function of the sample bias.

3.2.2. *The electronic structure of steps*

We now consider the electronic structure of the step itself [24]. In Fig. 11(A,B) we show dI/dV linescans across Au(111) and Ag(111) monoatomic steps. The sample bias is -0.12V for Ag(111) and -0.45V for Au(111), i.e. near or below the energy onsets of the corresponding surface states [9]. STM constant-current linescans indicating the profiles of the steps are shown as dashed curves. At these energies, the dI/dV linescans of both the Au and Ag samples show a maximum centered at the bottom of the step and a minimum near the top of the step. However, if one monitors the LDOS peak as a function of increasing sample bias (i.e. electron energy) as shown in Fig. 11(C), one observes that although its intensity and shape change, the position of the peak remains unchanged until the Fermi level is crossed, whereupon it moves towards the top of the step. We note that when the sample bias is negative, occupied states are probed, while at positive bias the density of unoccupied states is probed.

From the results in Fig. 11, we can conclude that, in relation to the unperturbed electron density of a terrace, there is an increased LDOS of occupied states near the bottom of the step, and a corresponding increase in the LDOS of unoccupied states near the top. Since, as we mentioned, this distribution is observed even below the onset of the surface state, we can conclude that the bulk electron density dominates the observed LDOS behavior. This dipole-like distribution is clearer below or near the onset of the surface state and becomes less well defined as higher energy, more diffuse surface state levels are probed.

The above observations are consistent with the the "electronic smoothing" effect proposed to operate at surface steps by Smoluchowski quite some time ago [25]. Smoluchowski argued that the electrons will not follow the sharp discontinuity of the atomic structure at a step because this would lead to a large electron kinetic energy, since the latter is proportional to the square of the gradient of the wavefunction. Instead, the electron density smooths out through a charge redistribution process which involves charge flow from the top of the step to the bottom [25]. The resulting dipolar charge distribution at steps can be viewed as a "step-edge" state, and it and the associated electric field are expected to affect chemisorption processes at steps.

3.2.3. *Adsorbed atoms and their interaction with surface states.*

Next we consider the electronic structure of adsorbed atoms and their effect on the surface states. Fig. 12 shows the tunneling spectrum of a sulfur atom at a Ag(111) surface. The spectrum shown is actually the difference between the spectrum (dI/dV vs. V) recorded directly over the S-atom and the spectrum obtained over a clean surface site. It shows that the S-atom leads to an enhancement in the LDOS below E_F with a peak at ~0.2 eV and a decrease in LDOS above E_F. This general behavior is in accord with the jellium calculations of Lang [26], which predict that adsorbed S will appear as a protrusion when occupied states are probed and as a hole in unoccupied states.

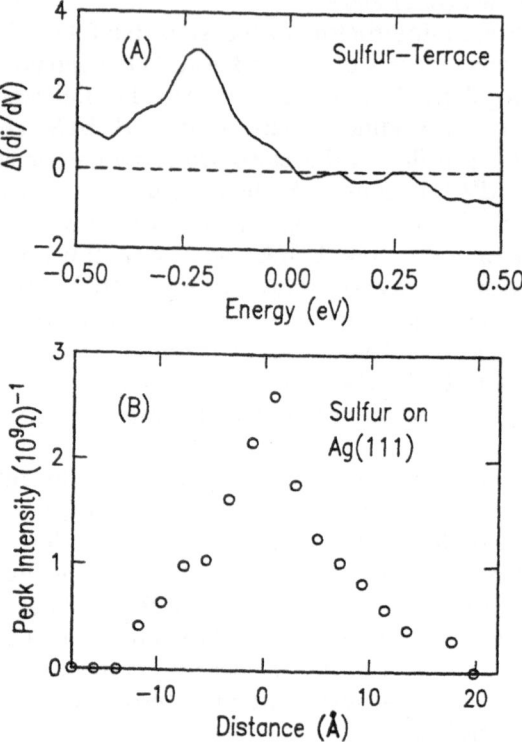

Fig. 12. (A) Difference between the tunneling spectrum recorded over an adsorbed sulfur atom on a Ag(111) surface and the spectrum obtained over a clean surface site. (B) Intensity of the dI/dV peak at $V_s = -0.2V$ as a function of the distance from the position of the sulfur atom.

However, there are rather large quantitative differences with the predictions of the jellium model. The latter predicts that the S 3p levels will be centered at ~2.5eV below E_F and have a width of ~2eV. The width we observe on Ag(111) is an order of magnitude narrower. In fact, this linewidth is significantly narrower than typical photoemission linewidths of adsorbate-induced states [27]. However, by considering the interaction of the S 3p levels with both the d and sp-bands of the substrate, one can understand qualitatively the position and the width of the S-induced peak [24]. The free S 3p levels (~10.4eV below the vacuum level) are in resonance with the Ag d-band, will interact with it and will be split into a set of bonding levels below the d-band, and a set of antibonding levels above it. Within each group the $3p_z$ and $3p_{xy}$ levels will be split in turn as a result of the bonding to the substrate. Finally, the S 3p levels will also hybridize and be broadened by the interaction with the Ag sp-valence band. This picture of the bonding was discussed in conjunction with the bonding of oxygen on nickel by Liebsch [28], and is supported by photoemission studies of S on Cu(100) [29]. Because of its energy the level probed with the STM should be one of the antibonding 3p levels. The fact that it is readily detected by the STM identifies it as a $3p_z$ level. We believe that the very small width of this level results from the fact that it lies at the edge of the Ag sp-gap at the L_2 point. Since there are no bulk sp states to hybridize with, the level remains sharp. It should also be noted that, unlike photoemission, the STM measurement is not subject to inhomogeneous broadening, and since we are dealing with an isolated adsorbate the peak is not broadened by dispersion effects. The difference with the jellium predictions can be attributed to the absence of the d-band and of the band-gap in the jellium model of the metal.

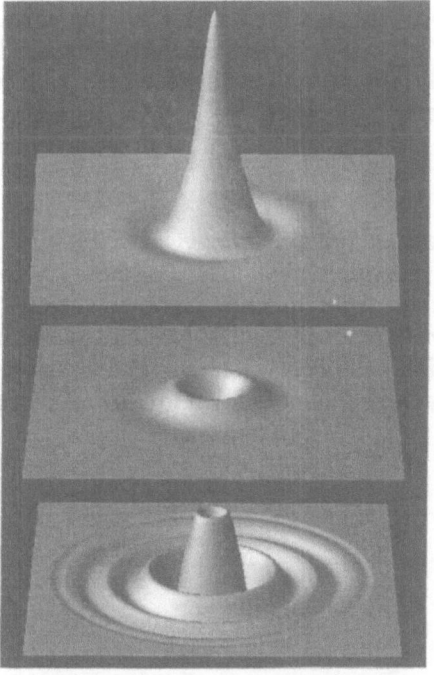

Fig. 13. Rotationally averaged dI/dV images of a sulfur atom on a Ag(111) surface as a function of sample bias. From top to bottom the bias is: -0.1V, +0.1V and +0.4V.

Now we can address the voltage dependence of the STM images. In Figure 13 we show rotationally-averaged 3D spectroscopic images of the S impurity at energies below (top) and above (middle and bottom) E_F. The peak in the occupied states arises from the $3p_z$ state as discussed above. The fact that the antibonding 3p-derived levels are occupied implies that the S atom attains a nearly closed 3p-shell. This explains why the S atom appears dark in positive bias (unoccupied states) images. The S 3d states are at much higher energy. Inverse-photoemission studies of S/Cu(100) place these 3d states at about 12eV above E_F [29]. Thus, the features near the origin in Fig. 13 can be accounted for on the basis of the intrinsic electronic structure of the adsorbate. The oscillations surrounding the S ion, on the other hand, are the energy-resolved Friedel-oscillations [30] which result from the screening of the ion by the surface state electrons. The energy (bias) dependence of the intensity of these oscillations has been discussed elsewhere [11]. The scattering and the screening descriptions of the oscillations are equivalent. The screening description, however, allows one to obtain information on the net charge on the adsorbate and thus gain insight on the nature of the chemisorption bond.

Finally, we note the large spatial extent of the S peak. By plotting the intensity of the peak at -0.2eV as a function of the distance from the center and correcting for the background, we find a Lorentzian-like profile with a fwhm of ~10 Å (Fig. 3b). This is much larger than the ionic radius of S^{2-} (~2 Å) and about a factor of two higher than that predicted by the jellium calculation [26]. It is likely that this large width reflects the hybridization of the adsorbate wavefunction with the nearly degenerate surface state, thus making the adsorbate-induced state very localized in energy, but spread out in space.

3.2.4. *Electron Confinement in Low-Dimensionality Metal Structures*

The 2DFEG system produced by semiconductor heterostructures can be patterned through lithographic techniques or by electric fields to give even lower dimensionality structures, i.e. 1D structures (quantum wires) and 0D structures (quantum dots). Since, as shown above, steps act as barriers for surface electrons, they can be used to form low dimensionality metal structures.

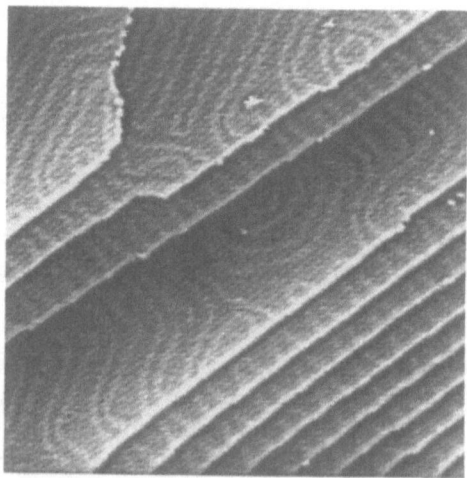

Fig. 14. STM image of a stepped Au(111) surface. The sample bias is +0.23 V. The pairs of lines seen on the terraces are due to the 22x+3 reconstruction of the Au(111) surface.

For example, one can form quasi-1D structures by creating narrow terraces whose width is of the order of the Fermi wavelength [31]. Stepped surfaces with variable terrace width can be produced by cutting a single crystal at a small angle with respect to a low-index crystal plane, or simply by the controlled annealing of thin films. Fig. 14 shows such a stepped Au(111) surface produced by annealing. 0D-like structures are generated by metal on metal evaporation and nucleation to form metal islands. Depending on the conditions, i.e. substrate temperature, surface coverage and annealing, islands with different sizes and shapes can be produced [31]. Both islands and narrow terraces are stable over a wide range of temperature. A different approach has been utilized by Crommie et al. [32] who arranged Fe atoms on Cu(111) at 4K to form closed structures, "quantum corrals", and thus generate confined states.

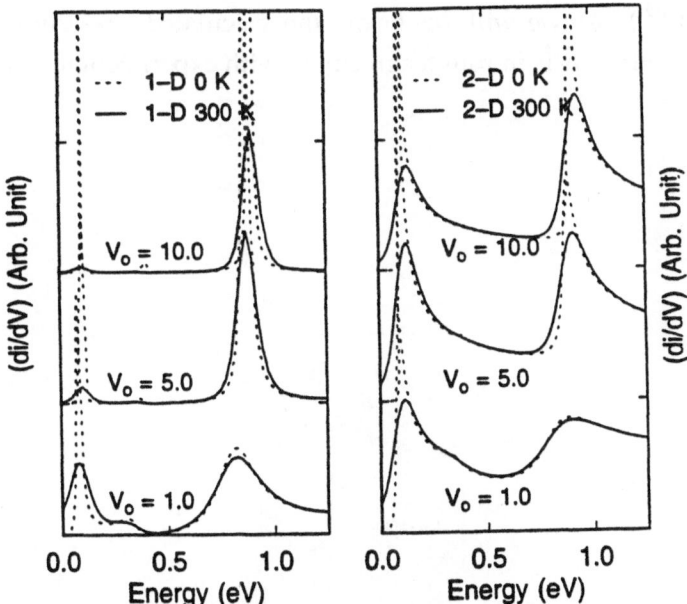

Fig. 15. Simulations of the spectra of confined electron states of a 36 Å -wide terrace as a function of the confining barrier V_0 in one- and two-dimensions (see text). Dotted lines and solid lines give results at 0K and 300K, respectively. The electron effective mass was taken to be that of the surface state of Au(111) (m^*=0.28).

Narrow terraces produce electron confinement in the direction normal to the step (x-direction), while electron motion parallel to the step (y-direction) is free. Thus, the resulting electron system is not trully 1D. The width and shape of the electronic spectra of such a system depend not only on the confining potentials but also on electron motion in the y-direction. To illustrate the influence of these factors and in order to compare with experiment we use a simple model in which the confinement between two steps is described by a hard wall on the upper side of step one and a variable delta function barrier, $V_O\delta(x)$, at the lower side of the other. We then calculate the probability density at the middle of the terrace as a function of energy for a 1D system (x only) and a 2D system (x and y) at both T=0K and 300K. Results for a 36 Å - wide terrace and parameters corresponding to the surface state of Au(111) are shown in Fig. 15. The strong dependence of the spectral shapes, widths, and intensities on the confining barrier height and on dimensionality is clearly evident. The spectra taken at the middle of the terrace show only the odd, n=1,3 states, while the n=2 state appears weakly only when the difference between the confining potentials is large. Most importantly, the 2D system is seen to exhibit a quasi-1D electronic behavior analogous to that seen in quantum wires. The asymmetric character of the spectra results from the fact that 1D DOS in the y-direction decreases with increasing energy ($\propto (E_0 - E)^{-1/2}$). As we will see below the calculated spectrum for a 2D system with V_0=1eV Å is in rough agreement with experimental results.

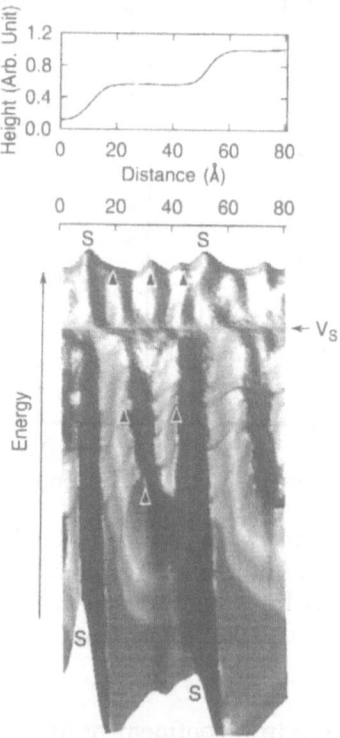

Fig. 16. Top: STM linescan across a narrow (36 Å) terrace of a stepped Au(111) surface. Bottom: Probability density distribution of the confined electron states of this terrace at 300K. This is a 3D map of dI/dV ($\propto$LDOS) as a function of the distance perpendicular to the steps, and of the sample voltage which varies from -0.47V (bottom), to +0.38V (top). Step-edge peaks are marked by S. The stabilization voltage is +0.23V.

In Fig. 16 we show experimental evidence of confinement at a 36 Å -wide Au(111) terrace at 300K. This is a 3D plot of the probability density (∝ dI/dV) of the confined states of the terrace, as a function of the distance perpendicular to the step and of the electron energy which increases from bottom to top. In the blue area at the bottom of the figure the energy is lower than the onset of the surface-state band. The structure seen in this case is due to the electronic structure of the steps (marked S), and to the corrugation due to the 22x√3. reconstruction of the Au(111) surface. After crossing into the surface band we observe the probability distribution of the nodeless n=1 standing-wave state formed between the two steps, this then changes into the distribution of the n=2 state (one node), and finally we observe the distribution of the n=3 state (two nodes). The evolution is gradual due to the broadening produced by the unconstrained motion in the direction parallel to the steps, the low confining barriers, temperature, and surface-to-bulk scattering.

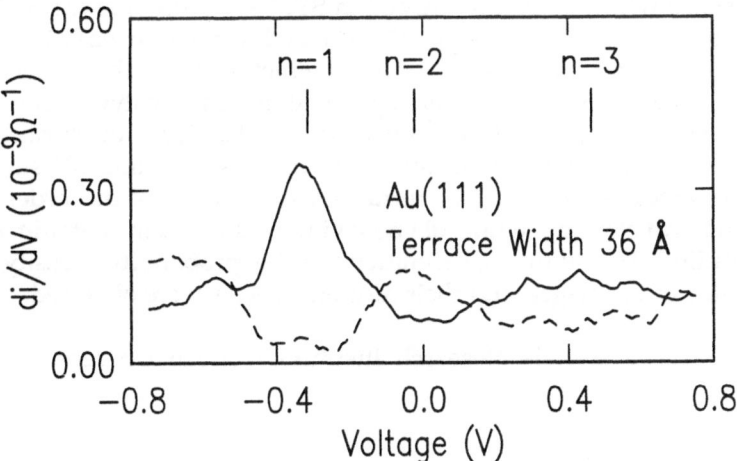

Fig. 17. Tunneling spectra of the confined states of a 36 Å -wide Au(111) terrace. The solid line gives the spectrum obtained with the tip placed over the center of the terrace, while the dotted line gives the spectrum with the tip placed at a distance equal to 1/4 of the terrace width from the step edge.

The formation of confined states at the narrow terrace is also reflected in the discrete nature of its spectra. Fig. 17 shows the tunneling spectra of the terrace. As we saw in the simulations in Fig. 15, the tunneling spectra are position-dependent. Because there is an approximate center-of-symmetry, only odd quantum number states are observed with the tip positioned at the center of the terrace (even n states have a node at the center). Finally, we note that 0D-electron confinement can also be observed at room temperature. We have observed such behavior in studies of silver islands produced by evaporation on top of a flat silver surface [31].

4. *Summary and Conclusions*

In conclusion, we have demonstrated two new applications of the STM. First, we utilized tip-sample point contacts to search for electrical transport through the dangling-bond surface state of Si(111)-7x7. The non-ohmic character and large spreading resistance of these nano-contacts, in conjunction with chemical quenching experiments, allowed us to prove the existance of such a transport channel. We also demonstrated that the electrical properties of individual nanostructures can be probed using tip-sample point contacts.

In the second part of the paper we showed that the interaction of surface state electrons of metals with individual surface features such as a step or an adsorbate can be studied using the STM. Electron scattering by these features leads to electron interference and the production of an oscillatory local-density-of-states which is readily imaged in STM spectroscopic maps. Results were presented for Au(111) and Ag(111) surfaces. Information was obtain about the electronic structure of the steps, their interaction with the surface state, bulk-surface state mixing, and corresponding issues involving individual adsorbates. The fact that surface steps act as barriers for surface state electrons was used to confine them and to form quasi-1D- and 0D-structures. Quasi-1D behavior is exhibited by terraces with widths of the order of the surface Fermi wavelength, while 0D behavior was found in metallic islands of similar radius. The spatial distributions of the probability density of the confined states were imaged and their discrete spectra were obtained.

* Present address: Materials Research Institute, Tohoku University, Sendai 980, Japan.

References

1. G. Binnig and H. Rohrer, Ch. Gerber, and E. Weibel, Phys. Rev. Lett. **49**, 571 (1982).
2. H.J. Gutherodt and R. Wiesendanger, Eds., Scanning Tunneling Microscopy, Vol. I, II, and III, (Springer-Verlag, Berlin, 1992); J.A. Stroscio and W.J. Kaiser, Eds., Scanning Tunneling Microscopy, (Acedemic Press, Boston, 1993).
3. K. Takayanagi, Y. Tanishiro, M. Takahashi, and S. Takahashi, J. Vac. Sci. Technol. A **3**, 1502 (1985).
4. D.E. Eastman, F.J. Himpsel, J.A. Knapp, and K.C. Pandey, Physics of Semiconductors, B.L.H. Wilson, ed. (Inst. of Physics, Bristol, 1978) p. 1059.
5. F.J. Himpsel and Th. Fauster, J. Vac. Sci. Technol. A **2**, 815 (1984).
6. R.J. Hamers, R.M. Tromp, and J.E. Demuth, Phys. Rev. Lett. **56**, 1972 (1986).
7. M. Henzler, in: Surface Physics of Materials, Vol. 1 (Academic Press, New York, 1975), and personal communication (Y.H.).

8. B.N.J. Persson and J.E. Demuth, Phys. Rev. B **30**, 5968 (1984); B.N.J. Persson, Phys. Rev. B **34**, 5916 (1986).

9. S.D. Kevan and R.H. Gaylord, Phys. Rev. B **36**, 5809 (1987).

10. L.C. Davis, M.P. Everson, R.C. Jaklevic and W. Shen, Phys. Rev. B **43**, 3821 (1991).

11. Y. Hasegawa and Ph. Avouris, Phys. Rev. Lett. **71**, 1071 (1993); Ph. Avouris et al., J. Vac. Sci. Technol. B **12**, 1447 (1994).

12. M.F. Crommie, C.P. Lutz, and D.M. Eigler, Nature **363**, 524 (1993).

13. Ph. Avouris and R. Wolkow, Phys. Rev. B **39**, 5071 (1989).

14. R. Landauer, I.B.M. J. Res. Develop.1, 223 (1957); J. Phys. Condens. Matter **1**, 8099 (1989).

15. S.M. Sze, Physics of Semiconductor Devices, 2nd ed. (Wiley, New York, 1981).

16. Ph. Avouris, I.-W. Lyo and Y. Hasegawa, J. Vac. Sci. Technol. A **11**, 1725 (1993).

17. F.J. Himpsel, G. Hollinger, and R.A. Pollack, Phys. Rev. B **28**, 7014 (1983).

18. T. Ando, A.B. Fowler, and F. Stern, Rev. Mod. Phys. **54**, 437 (1984).

19. J.J. Harris et al., Rep. Prog. Phys. **52**, 1217 (1989).

20. Ph. Avouris, Solid State Commun. **92**, 11 (1994).

21. J. Friedel, Adv. in Phys. **3**, 446 (1954); J. de Physique et le Radium, **17**, 27 (1956).

22. Y. Hasegawa and Ph. Avouris, Science **258**, 1763 (1992).

23. S.D. Kevan, and R.H. Gaylord, Phys. Rev. Lett. **57**, 2795 (1986).

24. Ph. Avouris, I.-W. Lyo, and P. Molinas-Mata, Chem. Phys. Lett. **240**, 423 (1995).

25. K. Smoluchowski, Phys. Rev. **60**, 61 (1941).

26. N.D. Lang, Phys. Rev. Lett. **56**, 1164 (1986).

27. E.W. Plummer and W. Eberhard, in Advances in Chemical Physics, vol. XLIX, I. Prigogine and S.A. Rice, editors (Wiley, New York, 1982).

28. A. Liebsch, Phys. Rev. B **17**, 1753 (1978).

29. G. Leschik, R. Courths and H. Wern, Surf. Sci. **294**, 355 (1993).

30. J. Friedel, Adv. in Phys. **3**, 446 (1954).

31. Ph. Avouris and I.-W. Lyo, Science **264**, 942 (1994).

32. M.F. Crommie, C.P. Lutz and D.M. Eigler, Science **262**, 2181 (1993).

SFFM AND SNOM OF HETEROGENEOUS MATERIALS

O. MARTI, E. WEILANDT, A. ROSA, J. STAUD, B. ZINK,
I. HÖRSCH, R. KUSCHE, O. KIRSCHENHOFER,
O. HOLLRICHER
Abteilung Experimentelle Physik
Universität Ulm
D-89069 Ulm
Germany

Abstract

It is often difficult to characterize and distinguish heterogeneous materials with domain sizes below 1 micrometer. The range between 1 nm and 1 μm lies in the crossover region where light scattering is applicable to the upper wavelength range and where x-ray and neutron scattering is strongest for the lower wavelength end. In addition the scattering methods require careful experimental procedures for a reconstruction of the real space structure. Scanning probe methods do provide a direct access to the surface properties in the range of interest. They are complementary to the recently developed x-ray imaging methods. In this paper we discuss the application of the scanning force and friction microscope and the near-field optical microscope to heterogeneous surfaces. The image formation process in both microscopes is different: the scanning force and friction microscope probes the nanomechanical properties of sample surfaces. The near-field optical microscope on the other hand measures the reflectivity, absorption or the fluorescence near the sample surface. As an application we discuss the imaging of polymer surfaces.

1. Introduction

Scanning probe microscopes are descendants of the scanning tunneling microscope (STM)[i]. The STM allowed for the first time to probe profiles of surfaces with atomic resolution. While other techniques like the transmission electron microscope or the field emission microscope do have similar resolution, they do not provide the intuitive understanding the STM does. This helped to spread the STM to many laboratories. However it was soon realized, that the imaging mechanisms were not all that easy to understand.

R. Rosei (ed.),
Chemical, Structural and Electronic Analysis of Heterogeneous Surfaces on Nanometer Scale, 25–41.
© 1997 *Kluwer Academic Publishers.*

The highly nonlinear interaction of the STM tip with the sample surface created enormous difficulties in the mathematical handling of the physical problems. While first applications of the STM[i] were limited to clean surfaces under well controlled vacuum conditions the need for more versatile imaging conditions arose. The first offsprings of the STM were the near field optical microscope (SNOM)[ii] and the scanning force microscope (SFM)[iii]. We will discuss SFM and SNOM and their application to the characterization of heterogeneous materials.

Scanning Force Microscopy is one of the most versatile microscopic techniques available today. Since its first appearance as a contact mode imaging microscope the technique has been refined to allow operation in air[iii,iv], in liquids[v], and under ultra high vacuum[vi] conditions in contact and non contact modes[vii]. The detection technique for forces has been improved to allow simultaneous measurements of forces. We will discuss selected applications which use this capability.

The near field optical microscope SNOM did have much a slower start than the SFM. However recent experiments have shown the capability for an analytical operation of the instrument[viii].

2. Resolution

If we want to compare scanning probe methods with other type of microscopes we need to get an understanding on how the resolution is achieved. Scanning probe microscopes are unique in that the sensing probe and the sample are in intimate contact. Therefore one can expect a highly nonlinear response to changing signal levels.

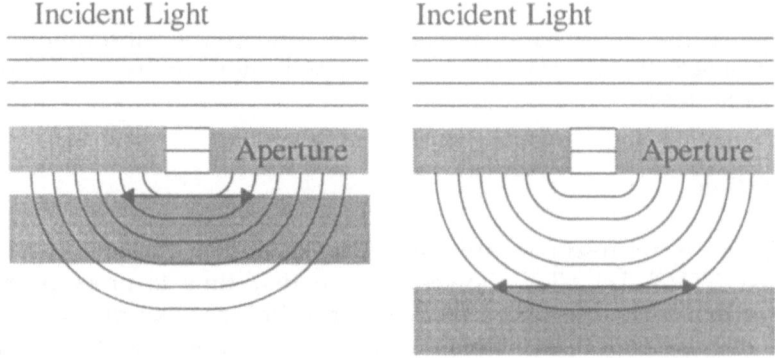

Fig. 1. Resolution given by apertures. The left image shows that the resolution of an aperture dominated microscope is better when the sample is closer to the aperture. The arrow shows the expected resolution.

In the case of a wave like interaction, as in the SNOM, an aperture determines the resolution. Figure 1 shows the effect of varying the distance between the aperture and the sample. The arrow in the picture shows, that for larger distances the resolution decreases. Observing that the intensity of a spherical wave depends on distance like

$$I(r) = \frac{I_0}{r^2} \tag{1}$$

and assuming that the size of the aperture is negligible compared to the distance of observation we can calculate the resolution. The aperture is assumed to be at a distance d from the sample.

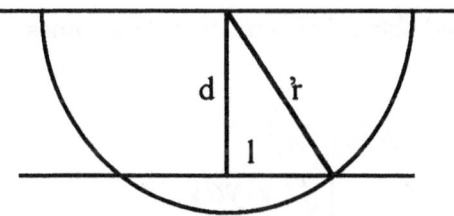

Fig. 2. Calculation of the aperture.

We define the resolution, somewhat arbitrarily, as the diameter $2l$ of the area which is illuminated with at least an intensity of $I(d)/2$ (see figure 2). r' is then given by

$$r' = \sqrt{2}\, d \tag{2}$$

We note that l is one side of a triangle with a 90° angle. Hence l becomes

$$l = \sqrt{r'^2 - d^2} = d \tag{3}$$

This means, that any time we have a wave like interaction the resolution of a scanned probe microscope is proportional to the distance between tip and sample. Besides light also acoustical waves do follow this law.

2.1 Dependence on interaction strength

Other scanned probe microscopes like the SFFM can not be understood with an aperture type tip. There the interaction is assumed to originate from the surfaces and the adjacent regions. This is not strictly true, but the approximation is best for interactions with a strong distance dependence. For a general interaction of which the distance dependence is given function $f(z)$ we can calculate the interaction to be

$$I(r) = \int_0^{2\pi} \int_0^{r_0} f(\sqrt{(R+d)^2 + r^2})\, r\, dr\, d\theta$$

$$= 2\pi \int_0^{r_0} f(\sqrt{(R+d)^2 + r^2})\, r\, dr$$

(4)

where R, d and r are defined in figure 3. We can now substitute $y = \sqrt{(R+d)^2 + r^2}$ and obtain from equation (4) becomes

$$I(r) = 2\pi \int_{R+d}^{\sqrt{(R+d)^2 + r_0^2}} f(y)\, y\, dy$$

(5)

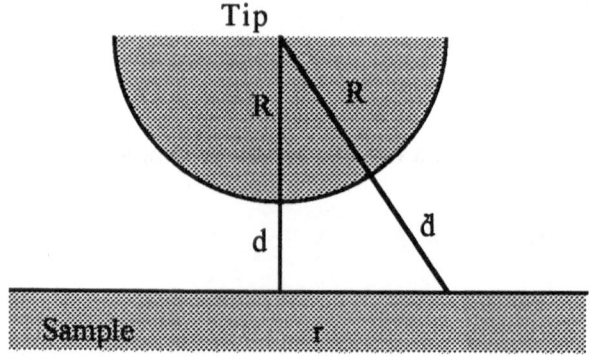

Fig. 3. Tip with a radius of curvature.

The resolution of a microscope can now be defined as the radius r_0 within which the fraction $A<1$ of the total interaction intensity. This means that we have to solve the equation

$$A = \frac{I(r_0)}{I(\infty)} = \frac{\displaystyle\int_{R+d}^{\sqrt{(R+d)^2+r_0^2}} f(y)ydy}{\displaystyle\int_{R+d}^{\infty} f(y)ydy} \tag{6}$$

For simple laws this equation can be solved. In scanning force microscopy one often uses inverse power laws. This results in interaction laws like the Lennard-Jones-potential. The effective distance is the one between the tip surface and the sample surface. With

$$f(y) = \frac{1}{(y-R)^n} \tag{7}$$

equation (6) becomes

$$A = \frac{\left(\dfrac{1}{d^{n-1}} - \dfrac{1}{\left((R+d)^2+r_0^2\right)^{n-1}}\right)}{\dfrac{1}{d^{n-1}}} \tag{8}$$

$$= 1 - \left(\frac{d}{(R+d)^2+r_0^2}\right)^{n-1}$$

And finally we obtain for the resolution

$$r_0 = \sqrt{\frac{d}{\sqrt[n-1]{1-A}} - (R+d)^2} \tag{9}$$

Equation (9) shows, that the resolution improves with steeper potentials.

2.2 Dependence on tip dimensions

It is instructive to first consider a SNOM in the so called illumination mode. There the sample is illuminated through a small aperture. Fig. 4a) shows a sketch of the geometry. If the tip is separated from the sample by a distance much smaller than the characteristic length of the interaction, the diameter A of the aperture will determine the resolution. This simple picture is only true, as long as the borders of the aperture can be considered perfect. For near field optical microscopes this is the case when the penetration depth of the light into the metal cladding is small compared to the diameter of the aperture. When this is not the case the complete interaction between the tip and the sample as well as the behavior of the metal cladding have to be taken into account.

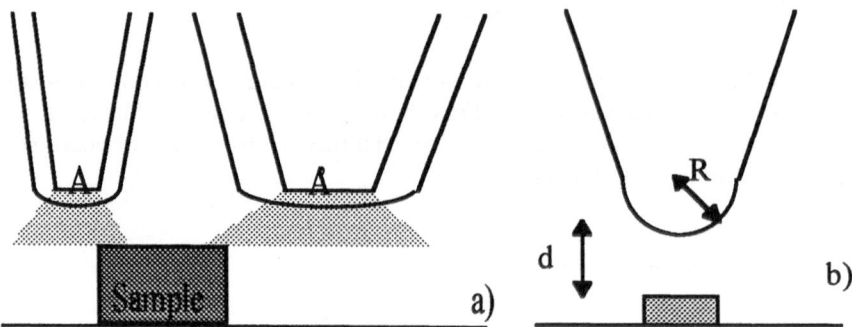

Fig. 4. The resolution of a scanning probe microscope is either determined by some aperture (part a)) in the case of the near field optical microscope or by the radius of curvature of the tip for an STM or SFFM (part b)).

2.3 Tip Effects while Imaging Large Features

The imaging of large features in a scanning probe microscope is dominated by the shape of the tip. Fig. 5 demonstrates that a blunt tip will broaden protruding features and will not be able to image down to the bottom of a narrow trench[ix].

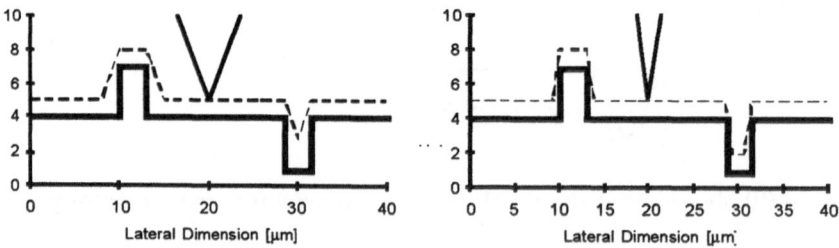

Fig. 5. Tip dependent resolution. The left side depicts how a fictional surface is imaged by a blunt tip. The left side depicts images of a sharp tip.

3. Scanning Friction Force Microscopy

The scanning force microscope[iii] images surfaces by using interaction forces. Forces are measured by springs, usually cantilevered leave springs[x]. A force applied to the cantilevered spring (usually named cantilever) deflects the end. Hence the measurement of forces reduces to a measurement of small distance changes or to a measurement of small changes in angular orientation. The first method we will discuss, the optical lever deflection, is indeed sensitive to the orientation of the end of the cantilever. A second method in use is optical interferometry. Interferometry measures distances. Not very widely used anymore are force microscopes using a tunneling junction to the cantilever. We will not discuss them here.

3.1 Optical Lever Detection System.

The optical lever detection system is a simple yet elegant way to detect normal and lateral force signals simultaneously[x,xi]. A second advantage is the fact that it is a remote detection system.

Light from a laser diode or from a super luminescent diode is focused on the end of the cantilever. The reflected light is directed onto a quadrant diode which measures the direction of the light beam. A Gaussian light beam far from its waist is characterized by an opening angle β. The deflection of the light beam by the cantilever surface tilted by an angle α is 2α. The intensity on the detector then shifts to the side by the product of 2α and the separation between the detector and the cantilever. The readout electronics calculates the difference of the photo currents. The photo currents in turn are proportional to the intensity incident on the diode.

The output signal is hence proportional to the change in intensity on the segments

$$I_{sig} \propto 4\frac{\alpha}{\beta}I_{tot} .$$

(10)

For a Gaussian beam the resulting output signal as a function of the deflection angle is dispersion like. Equation (10) shows that the sensitivity can be increased by increasing the intensity of the light beam I_{tot} or by decreasing the divergence of the laser beam. The upper bound of the intensity of the light I_{tot} is given by saturation effects on the photo diode. If we decrease the divergence of a laser beam we automatically increase the beam waist. If the beam waist becomes larger than the width of the cantilever we start to get diffraction. Diffraction sets a lower bound on the divergence angle. Hence one can calculate the optimal beam waist w_{opt} and the optimal divergence angle β[8]

$$w_{opt} \approx 0.36b$$

$$\theta_{opt} \approx 0.89\frac{\lambda}{b}$$

(11)

where β is the width of the cantilever and λ is the wavelength of the light. The optimal sensitivity of the optical lever then becomes

$$\varepsilon[mW/rad] = 1.8\frac{b}{\lambda}I_{tot}[mW] \qquad (12)$$

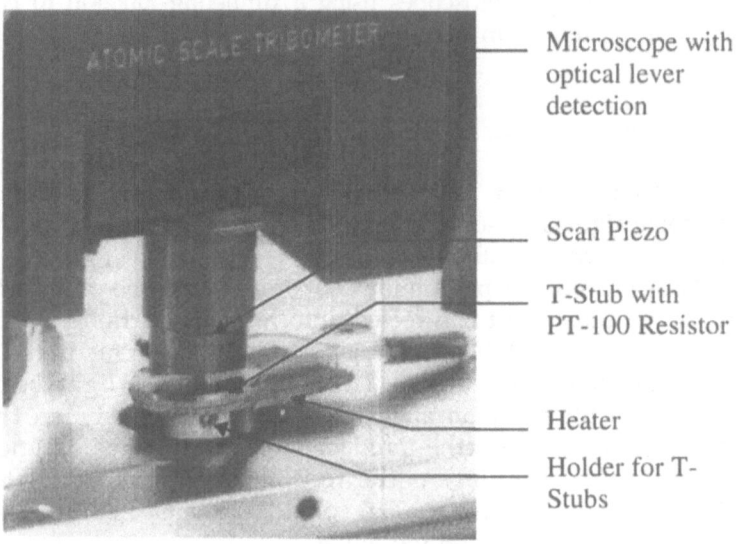

Microscope with optical lever detection

Scan Piezo

T-Stub with PT-100 Resistor

Heater

Holder for T-Stubs

Fig. 6. Construction of a heating stage for a commercial scanning force microscope[xii].

The angular sensitivity optical lever can be measured by introducing a parallel plate into the beam. A tilt of the parallel plate results in a displacement of the beam, mimicking an angular deflection.

3.2 Polymer Samples

Polymers are in an important class of materials for industrial use. Intense research efforts are under way to create materials tailored to specific tasks and needs. The surface properties of polymers are important for their performance. The scanning force and friction microscope now allows to probe mechanically polymer surfaces. The versatility of the SFFM enables experiments not only under ambient conditions but also at elevated temperatures.

3.2.1 PMMA, Temperature Dependent Imaging

A first example is the investigation of PMMA (Polymethylmethacrylate). Because the physics of PMMA has been extensively studied in the literature it is a good test candidate for the SFFM[xiii].

Fig. 6 displays the construction of a heating stage for a commercial scanning force microscope. The design idea was to minimize the heated volume. The sample is mounted on a standard aluminum T-stub. A hole drilled in the sample holder dish from the side holds a PT-100 resistor. A temperature controller supplies the heating power to the heater foil. The heating stage operates from room temperature to 200 °C. Temperature measurements on the microscope have shown that the safe operating area is up to 150 °C: up to this temperature the heating of the microscope is negligible.

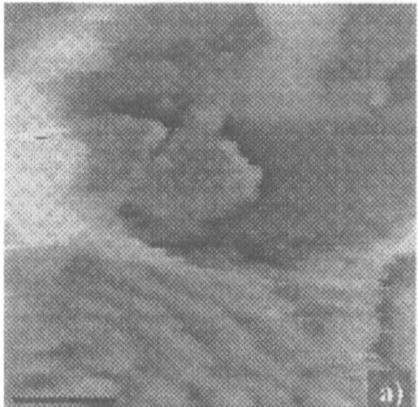 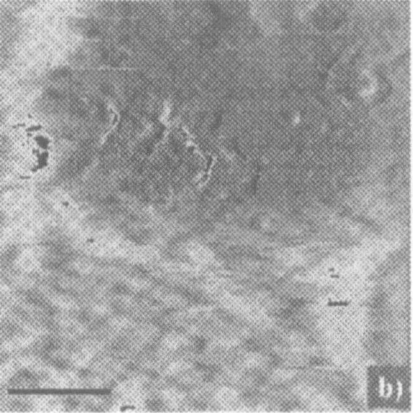

Fig. 7. Scanning Force and Friction Microscope image of PMMA at 120.2 °C. The bar indicates a length of 1 μm. The height scale in the topography image a) is 260 nm, image b) is the friction force image.

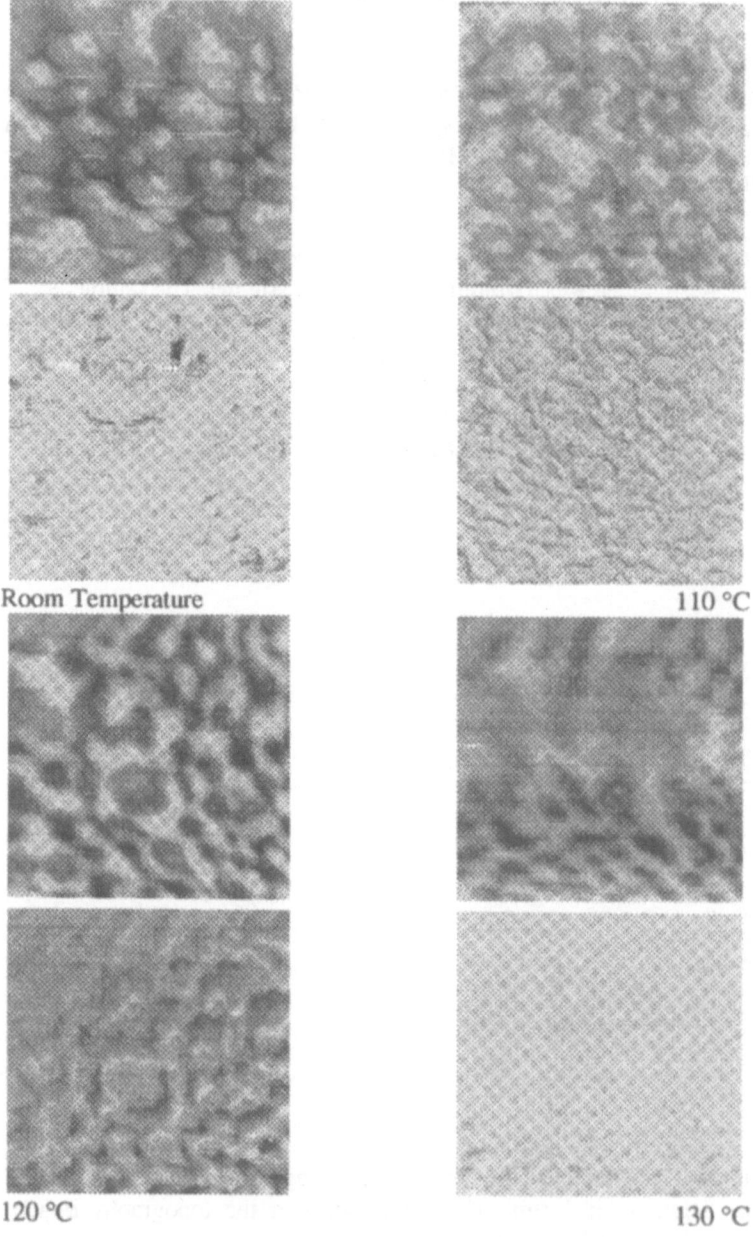

Fig. 8. PMMA at different temperatures. The top row and the 3rd row display the topography of the PMMA surface, the 2nd row and the bottom row shows the local stiffness. The data was measured with the pulsed force mode. Each image displays an area of 4 x 4 μm^2. The scale for the topography is 32 nm

Fig. 7 shows a measurement of PMMA at a temperature of 120.2 °C. The PMMA was spin cast onto a glass substrate. Its molecular weight was 4.41×10^5 g/mol and its dispersity 2.1. The glass temperature of the sample material was about to 100 °C Hence Fig. 7 was measured above the glass temperature. The PMMA covers the lower part of the image The wavy structure of the PMMA-surface is induced by the force microscope tip[xiv] . The interaction forces together with the scanning of the tip serve to build up these waves. The microscope itself is operating correctly, as can be seen from the upper part of the image. Small gold spheres have been deposited on the substrate prior to the deposition of the PMMA. The gold spheres are imaged in the PMMA-free zone at their correct dimension. To overcome the problem of the friction and the adhesion of the PMMA we used a sampling technique which we call the pulsed force mode[xii]. The tip of the force microscope is periodically lowered down to the sample surface until a preset force is reached. The tip is then withdrawn. measuring a point on the force-distance curve allows us to determine not only the topography but also the local stiffness. The height of the sample surface is determined more than 5000 times a second. Since the surface position is probed at a relatively high frequency the surface is dynamically stiffer. Fig. 8 shows an example of such a measurement. The sample has the same composition as the one discussed in Fig. 7. The glass temperature is about 100 °C. Even though the PMMA is in a liquid like state at 130 °C the topography can be faithfully reproduced. The network structure of the PMMA surface evident at all temperatures is due to the preparation process. The spin cast sample has not been cured: hence it is not in a thermodynamic equilibrium.

3.2.2 Variation of External Parameters

Fig. 9. Effect of the stretching of a filled rubber. The sample has been stretched by 200 % in the direction indicated by the arrow. The side of the image has a length of 3.5 μm. The left image shows the topography, the central one the local stiffness and the right one the adhesion.

An advantage of a stand-alone scanning force microscope[xv] is the easiness with which additional experiments can be performed on the sample. An example is the investigation of the surface morphology of filled rubbers[xvi]. The structural strength and the mechanical parameters of such a rubber sample are intimately connected to the interaction of the rubber molecules with the filler particles. The scanning force and friction microscope can now be used to probe the surface of a rubber sample during the stretching experiment. Fig. 9 shows a typical result, on the left side the topography and in the center the local stiffness. The right side shows a measurement of the local adhesion[xvii]. The local adhesion is determined by a measurement of the force necessary to detach the SFFM tip from the sample surface. While the local stiffness measurement indicates the presence of a filler particle in the sample, the local adhesion might be able to distinguish between bare filler particles and filler particles covered by a thin polymer film. Adhesion thus has the potential to generate a signal related to the sample composition.

4. Near Field Optical Microscopy

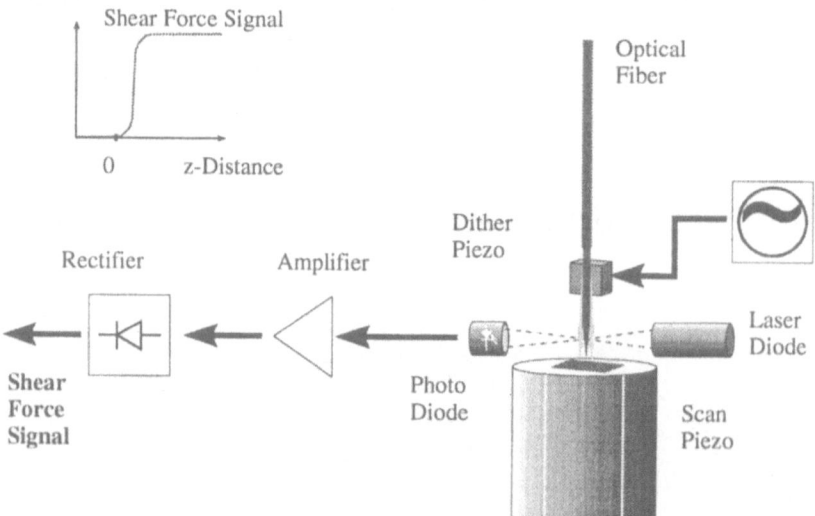

Fig. 10. Schematic setup of a near field optical microscope. including a shear force control of the tip-sample distance.

Complementary to the scanning force and friction microscope is the scanning near field optical microscope (SNOM)[ii]. This microscope allows to measure optical properties, reflectivity, transmissivity, fluorescence, luminescence. These measurements can be static or time resolved. The ultimate time resolution is given by the pulse length one can feed into a near field optical microscope. In this paragraph we will discuss two simple applications of the SNOM.

4.1 Distance control by Force Microscopy

It can be shown that the resolution of an aperture SNOM is not only determined by the aperture of the tip (or the diameter of the light source) but also by the distance between the tip and the sample (see also 0). The SNOM consists of the usual components of a scanning probe microscope: the piezo scanner, a tip detecting or inducing an interaction and a feedback control system. The shear force[xviii] is measured by vibrating a fiber parallel to a sample surface and measuring the instantaneous amplitude. The vibration is damped by a yet undetermined interaction. The vibration amplitude is measured by determining the varying amount of shading by the fiber of a laser beam. The position of the sample is adjusted in such a way as to keep the damping constant. Typically there exists a gap of a few nanometers to a few 10 nanometers between the tip and the sample surface. Hence a near field optical microscope provides a non-contact force microscopy measurement of the sample topography.

4.2 Transmission mode

As an example we show a measurement of a test pattern[xix]. Fig. 11 shows the topography on the left side and the transmission image on the right side. The experiment has been performed with an uncoated fiber, pulled in a commercial fiber puller[xx] to a radius of curvature of less than 100 nm. The test pattern contains lines, dots and letters with various sizes. The structure is made from chromium, and written by electron lithography. From the topography image in Fig. 11 it is obvious that the shear force detection imaged with a multiple tip. The different tips are of apparent equal strength in the topography. The right side of Fig. 11 shows a simultaneous transmission measurement. The resolution of the transmission mode is slightly better than 1 μm.

Fig. 11. Transmission image of a test pattern. The left side of the image shows the topography, the right side the transmission measured with an uncoated fiber and a wavelength of 488 nm.

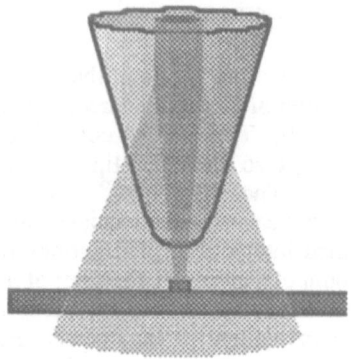

Fig. 12. An artist's sketch on the light path in near field optical microscope using uncoated fibers.

Fig. 12 shows an artist's sketch of a qualitative understanding of the origin of the resolution. Light is guided in the fiber core of a fiber. When the diameter of the fiber is reduced to form a tip the core can no longer guide the light. The glass-air boundary will serve as a total reflection surface until the diameter of the glass becomes less than about a wavelength. Then the light emerges from the fiber and illuminates the sample. The distance between the fiber and the apex of the tip is held constant. Hence the resolution of the transmission image is significantly worse than in the topography.

4.3 Reflection contrast

If the light reflected back into the fiber is measured instead of the transmitted light the experiment shows that the resolution is significantly improved (Fig. 13). Using the sketch in Fig. 12 we see that most of the reflected light can not couple back into the fiber.

Only a small fraction of the illuminated area reflects the light back into the fiber. This area is significantly smaller than the area involved in forming the topography contrast through the shear force. This can be seen by comparing the left image (topography) of Fig. 13 with the right one (reflection contrast). The topography shows clear evidence of multiple tips. However the reflection image is imaging one of the apparent protrusions. This is most obvious at the lower right side of the image. Using the resolved letters one can estimate the resolution of the microscope to 100 nm or better. Hence in reflection mode a good resolution can be achieved even though the aperture is large. A similar effect is known when using multi-photon spectroscopy in confocal microscopy[xxi]: the resolution increases because a higher power of the electric field of the light contributes to the image formation. In the reflection mode imaging of by a uncoated fiber we use the aperture twice: the apparent aperture becomes smaller.

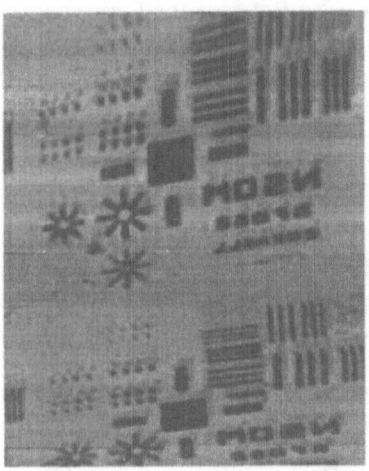 

Fig. 13. Reflection contrast of the same test pattern[xix] as in Fig. 13. The left side of the image shows the topography, the right side the transmission measured with an uncoated fiber and a wavelength of 488 nm. The large square has a dimension of 2 μm by 2 μm. The smallest readable letters have a size of 500 nm.

5. Conclusions

In this paper we have discussed some problems of the image formation in a scanning probe microscope. The resolution of a scanning probe microscope depends critically on the interaction of the tip and the sample as well as the shape of the tip. The steeper the interaction potential is the better the resolution of the microscope. The smaller the radius of curvature of the tip the better the resolution.

We have shown that the scanning force and friction microscope is a useful tool to investigate polymer surfaces. The sample holder of a standalone SFM has been modified to include a heating stage. This heating stage allows investigations in temperature range from room temperature to 150 °C. As an example we have discussed the images of PMMA. The same SFM has also been used to investigate the surface morphology of filled rubber samples. The measurement of adhesion gives additional information on the material contents of the sample.

Near field optical microscopes are an ideal complementary tool to get optical information from very small length scales. Superresolution reflection microscopy, transmission microscopy, fluorescence and luminescence microscopy is possible together with a non-contact type force microscopy of the sample topography.

Scanning probe microscopes are a useful complement of the photon or electron based imaging or scattering techniques. Scanning force microscopes provide otherwise not that easy to determine quantities like the local stiffness and

40

the local adhesion. The unique power of a near field optical microscope is to give simultaneous measurements of a surface topography and of optical properties. These may include reflectivity, fluorescence or luminescence or time resolved data[xxii].

6. Acknowledgments

We would like to thank G. Krausch, W. Wilke, J. Mlynek, M. Pietralla, B. Heise and G.-I. Asbach for discussions. Electronic circuits instrumental for the success of the experiments were built by G. Volswinkler. The test sample was kindly provided by R. Tiberio, J. Cline, G. Valaskovicz (Cornell) and G. Krausch (Konstanz). The authors are grateful for the financial support by the German Science Foundation (SFB 239) and the German Ministry of Education and Technology (BMBF). One of us (A. R.) gratefully acknowledges a grant from the Friedrich-Ebert-Stiftung.

7. References

[i] Binnig, G., Rohrer, H., Gerber, Ch., and Weibel, E. (1982) Phys. Rev. Lett. 49, 57;

[ii] Pohl, D.W., Denk, W. and Lanz, M. (1984) Appl. Phys. Lett. 44, 651; Dürig, U., Pohl, D.W., and Rohner, F. (1986) J. Appl. Phys. 59, 3318.

[iii] Binnig, G., Quate, C.F., and Gerber, Ch. (1986) Phys. Rev. Lett. 56, 930.

[iv] Albrecht, T.R., and Quate, C.F. (1987) J. Appl. Phys. 62, 2599.

[v] Marti, O., Drake, B., and Hansma, P.K. (1987) Appl. Phys. Lett. 51, 484.

[vi] Howald, L., Meyer, E., Lüthi, R., Haefke, H., Overney, R., Rudin, H., and Güntherodt, H.-J. (1993) Appl. Phys. Lett. 63, 117; Howald, L., Haefke, H., Lüthi, R., Meyer, E., Gerth, G., Rudin, H., and Güntherodt, H.-J. (1993) Phys. Rev. B.49, 5651.

[vii] Martin, Y., Williams, C.C., and Wickramasinghe, H.K. (1987) J. Appl. Phys. 61, 4723.

[viii] For an overview see: „Photons and Local Probes", eds. Marti, O. and Möller, R. (1995) NATO ASI Series E, in press.

[ix] O. Marti, „SXM: An Introduction", in „STM and SFM in Biology", O. Marti and M. Amrein, eds, Academic Press San Diego, (1993).

[x] Blanc, N., Brugger, J., and de Rooij, N.F. in „Forces in Scanned Probe Methods", Güntherodt, H.-J., Anselmetti, D., and Meyer, E. (1995) NATO

ASI E286 (Kluwer Scientific Publishers, Dordrecht), 79 and references therein.

[xi] Marti, O., Colchero, J., and Mlynek, J. (1990) Nanotechnology 1, 141; Güntherodt, H.-J., Anselmetti, D., and Meyer, E. (1995) NATO ASI Series E286 (Kluwer Scientific Publishers, Dordrecht)

[xii] Atomic Scale Tribometer, CSEM, Neuchâtel, Switzerland.

[xiii] Young, R.J. and Lovell, P.A. (1991) „Introduction to Polymers", Chapman & Hall (London). Material „Plexiglas" 5N from Röhm, Darmstadt, Germany.

[xiv] Radmacher, M., Fritz, M., and Hansma, P.K. submitted to Biophys. J., Radmacher, M., private communication.

[xv] Hipp, M., Bielefeldt, H., Colchero, J., Marti, O., and Mlynek, J.(1992) Ultramicroscopy 42-44, 1498

[xvi] Kilian, H.G., Schenk, H., and Wolff, S., (1987) Coloid &Polymer Sci. 265, 410. Material from Degussa, Wesseling, Germany.

[xvii] van der Werf, K.O., Putman, C.A.J., de Grooth, B.G., and Greve, J. (1994) Appl. Phys. Lett. 65, 1195; Radmacher, M., Fritz, M., Cleveland, J., Walters, D., and Hansma, P.K. (1994) Langmuir 10, 3809.

[xviii] Yang, P.C., Chen, Y., and Vaez-Iravani, M. (1992) J. Appl. Phys. 71, 2499; Betzig, E., Finn, P.L., and Weiner, J.S. (1992) Appl. Phys. Lett. 60, 2484.

[xix] R. Tiberio, J. Cline and G. Valaskovicz, National Nanofabrication Facility, Cornell University, USA.

[xx] Model P-2000 fiber puller, Sutter Instrument Company, Novato, CA, USA

[xxi] Stelzer, E.H.K., Hell, S., Lindek, S., Stricker, R., Pick, R., Storz, C., Ritter, G., and Salmon, N. (1993) Optics Comm. 104, 223.

[xxii] Pohl, D.W. and Courjon, D., eds. (1993) „Near Field Optics" NATO ASI Series E 242 (Kluwer Scientific Publishers, Dordrecht)

SYNCHROTRON RADIATION SPECTROMICROSCOPY: OPPORTUNITIES, LIMITATIONS AND DATA TAKING STRATEGIES

G. MARGARITONDO
Institut de Physique Appliquée, Ecole Polytechnique Fédérale
CH-1015 Lausanne, Switzerland.

1. Introduction

We discuss one of the most important yet often overlooked problems of today's experimental microanalysis: the need for careful data taking strategy planning. This problem is particularly relevant for the new class of synchrotron-radiation techniques known as *spectromicroscopies* -- which combine the analytical power of established spectroscopies with high lateral resolution.[1] These techniques are becoming extremely important with the advent of the new, ultrabright synchrotron sources of soft-x-rays: ELETTRA in Trieste, the Advanced Light Source in Berkeley, the SRRC in Hsinchu-Taiwan, and the Pohang source in Korea.

Microscopy and spectromicroscopy are indeed among the best ways to exploit the unprecedented levels of brightness of these new sources. This, however, may lead to a often unforeseen problems: an excessive amount of data which becomes impossible to process, or a waste of beamtime, which is very expensive for a sophisticated synchrotron source.

This problem has been analyzed in detail in Ref. 2. Here, we would like to use the results of Ref. 2 and draw some general strategic conclusions that might be useful for the many colleagues interested in using these new and potentially very powerful techniques.

43

R. Rosei (ed.),
Chemical, Structural and Electronic Analysis of Heterogeneous Surfaces on Nanometer Scale, 43–52.
© 1997 *Kluwer Academic Publishers.*

2. Some General Considerations

In order to put the discussion on a concrete ground, we will discuss here a specific photoemission spectromicroscopy experiment. Figure 1 shows a microimage created [3] with the photoemission spectromicroscope MAXIMUM at the Wisconsin Synchrotron Radiation Center. The image illustrates a Ge overlayer on a cleaved GaSe substrate. The image was created by focusing the soft-x-ray beam generated by an undulator, then scanning the sample with respect to the focal point while collecting photoelectrons at a fixed kinetic energy with an electron analyzer.

Although features are readily evident in the image, it is not clear *a priori* which is the mechanism that causes them. Changes in the photoelectron current from point to point can in fact be caused by a number of factors, such as:

- Topographic features, coupled to the anisotropic angular distribution of photoelectrons.

- Changes in the chemical composition that affect the core-level intensity of the corresponding elements.

- Changes in the chemical status of one or more elements, that affect the energy positions of the corresponding core levels.

- Changes in the local electrostatic shift with respect to the Fermi level, which could be due to diverse causes such as charging, band bending, etc.

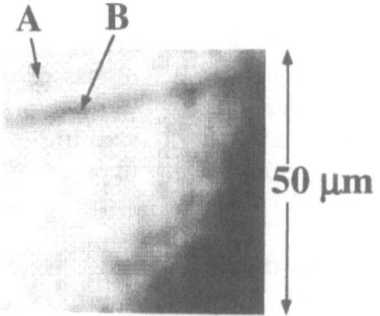

GaSe + 0.5 ML Ge

Figure 1. Photoelectron micrograph (Ref. [3]) of a Ge-covered GaSe substrate. The contrast mechanism is discussed in the text.

In the specific case of Fig. 1, it was found that the contrast between points A and B is related to a change in the band bending of the (semiconducting) GaSe substrate. In turn, as discussed in detail in Ref. 3, this change reveals a change in the band lineup of GaSe and Ge between the two points. The detection of such band lineup changes was indeed the main objective of the experiment of Ref. 3. Assume now that one would like to continue the experiment, taking additional data beyond the detection of the band lineup changes. As in any type of experimental work, it is quite important to clearly identify what are the objectives of the study, i.e., what is the information that one seeks to obtain. Then, it is equally important to take data in the right amount and of the right quality to obtain the desired information: not less, in order to avoid wasting an entire experiment (for which additional synchrotron beamtime might not become available), and not more, to avoid wasting beamtime which is a precious resource.

Assume, for example, that one decides to check if all core level peaks for the points of the upper dark line in Fig. 1 have the same energy positions. One could adopt a brute force strategy, by taking spectra for each and every point in the micrograph, then extracting the spectra for selected points in the dark line. This, however, could be a costly mistake: suppose that the time per spectrum is 5 minutes, and that one takes three core-level spectra (one for each element) per point; the total data taking time for the 100 ¥ 100 pixels in the micrograph is 9 ¥ 10^6 s, or 2,500 hours -- which at the current cost of synchrotron radiation undulator beamtime is equivalent to more than one million dollars (and would be impossible to obtain anyway). Of the spectra hypothetically obtained at such a high cost, one would then throw away more than 90%.

This examples dramatically illustrates the importance of the problem: the transition from spectroscopy to spectromicroscopy implies moving from one dimension (the energy axis) to three dimensions (energy and two surface coordinates). Thus, every mistake in data taking strategy is enormously amplified, and may lead to disaster.

This crucial point can be illustrated by a variety of possible mistakes. For example: can one be sure that one is not using excessive resolution? Higher resolution almost always means lower signal per unit time and therefore longer data taking time. A simple factor of two in the above experiment would bring the already prohibitive cost to more than two million dollars! We are thus forced to conclude that the data taking strategy, which is often neglected for simple spectroscopy, becomes an absolute necessity in spectromicroscopy.

3. The Question of Resolution

There are in fact many elements in the data taking strategy, whose analysis is often similar for different elements. We cannot analyze here more than a handful of examples, and we will begin with the very old issue of resolution: how much is resolution is "enough" for taking a given spectrum? And how much is "excessive" resolution?

Consider a simple photoemission spectrum, *i.e.*, a series of detected photoelectron intensities I as a function of a dimensionless energy parameter x. Suppose for simplicity that the spectrum is a Gaussian peak of width σ

$$I(x) = I_0 \exp[-(x-x_0)^2/\sigma^2] \; ; \tag{1}$$

the information content of the spectrum is given by the (missing) information entropy:

$$S = -k \int_{-\infty}^{\infty} [I(x)/I_0] \ln[I(x)/I_0] dx \quad + \text{constant} , \tag{2}$$

which in the case of the Gaussian peak of Eq. 1 becomes:

$$S = k \ln(\sigma) = \frac{k}{2} \ln(\sigma_R^2 + \sigma_I^2) + \text{constant} , \tag{3}$$

where σ_R and σ_I are the "real" (intrinsic) bandwidth, and the instrumental broadening. The meaning of Eq. 3 is apparently straightforward: the information entropy decreases, thus the information carried by the spectrum increases, if the bandwidth decreases. In particular, the information increases if the instrumental broadening decreases.

One should not, however, rush to the conclusion that one must bring the instrumental resolution to the limit. In order to increase the resolution, one often has to close some slits somewhere or do something else that decreases the signal-to-noise ratio. This in turn decreases the information content of the spectrum: there exists somewhere an optimum compromise between signal level and resolution, as all good practitioners of spectroscopy know.

In order to analyze this point, note that one does not need an infinitely narrow peak to determine with infinite accuracy its center x_0: if the lineshape is known (Gaussian in our example), then the accuracy limit in determining x_0 is determined by the signal-to-noise ratio. With infinite signal-to-noise ratio, a fitting procedure would determine x_0 with infinite accuracy. Conversely, a finite signal-to-noise ratio gives finite accuracy. The most important information content parameter, therefore, is not the simple information entropy of Eqs. 1 and 3, but the "maximum extractable information" entropy, which corresponds to the maximum possible amount of information extractable from a spectrum with data processing such as least-square fitting or instrumental broadening deconvolution.[2] One can show that, for stochastic noise, the "maximum extractable information" entropy, in the case of Eq. 3 is [2]:

$$S_L = \frac{k}{2} \ln[(\sigma_R{}^2 + \sigma_I{}^2)/I_0] + \text{constant} ,$$ (4)

where the difference with respect to Eq. 3 is given by the factor I_0, in turn related to the signal-to-noise level.

If one tries to decrease this entropy by decreasing the instrumental broadening s_I, the gain is offset by the decrease in signal level, I_0. Assuming that the two parameters are linearly related to each other, as it is often the case for practical spectrometers, we have:

$$I_0 = A\sigma_I$$ (5)

(where the coefficient A depends on several factors such as the slit widths and, most importantly, the accumulation time per data point, t), and Eq. 4 becomes:

$$S_L = \frac{k}{2} \ln[(\sigma_R{}^2 + \sigma_I{}^2)/\sigma_I] - \frac{k}{2} \ln(A) + \text{constant} .$$ (6)

The meaning of this equation is both simple and interesting: the maximum (extractable) information does not continuously increase by decreasing the instrumental broadening, but reaches an optimum (minimum entropy) for $\sigma_R = \sigma_I$, when the intrinsic and instrumental broadening are equivalent. Any additional attempt (Fig. 2) to decrease the instrumental broadening is technically an overkill.

The decrease in the maximum (extractable) information content due to such an overkill could be compensated by increasing the magnitude of the proportionality coefficient A. This could be accomplished by increasing the accumulation time per data point, τ, but such is precisely the mechanism that leads to a long total data taking time: the overkill resolution causes an expensive waste of time.

48

4. From Spectroscopy to Spectromicroscopy

The previous analysis can be easily extended from spectroscopy to microscopy. The formalism is indeed similar. If one consider an one-dimensional "image", the "Gaussian peak" becomes a "dot" with a given intrinsic size broadened by instrumental (space) broadening. Thus, for a one-dimensional "image" the conclusions are the same as for the above one-dimensional spectrum: by pushing the instrumental broadening below the intrinsic width of the feature - the dot in this case - one falls in the overkill regime.

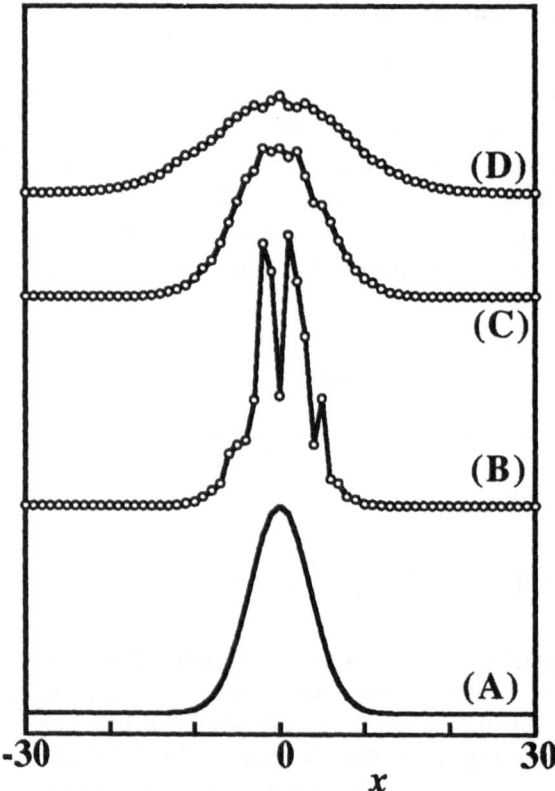

Figure 2. Optimization of the maximum (extractable) information content of a spectrum: (A) the "ideal" spectrum with no instrumental broadening and only the intrinsic broadening σ_R; (B)-(D): "real" spectra with instrumental broadening $\sigma_R = \sigma_I/3, \sigma_I$ and $2\sigma_I$, and the corresponding levels of noise. Spectrum (C) corresponds to the optimum information content determined by the interplay of broadening and noise.

For a two-dimensional image, the analysis is again similar. Consider, for example, a circularly symmetric Gaussian dot. One can easily demonstrate that the "maximum extractable information" entropy is again given by Eq. 6 if one assumes that Eq. 5 is isotropically valid. Therefore, the condition for optimum (extractable) information is again that the instrumental broadening equals the intrinsic broadening. The similar case of an infinitely sharp dot is illustrated by Fig. 3.

The impact of a mistake in the data taking strategy does increase on moving from one-dimensional images to two-dimensional images. The reason is clear from Eq. 6: suppose that one tries, for example, to correct an overkill in the instrumental broadening, which increases the first term, by an increase in the magnitude of the A-factor. This can be accomplished with an increase in the accumulation time per data point, τ. For a line image with N pixels, this increases the total data taking time following an Nt law. For a two-dimensional image with N x N pixels, the increase in the total data taking time follows instead a quadratic $N^2\tau$. The practical and financial difference can be dramatic.

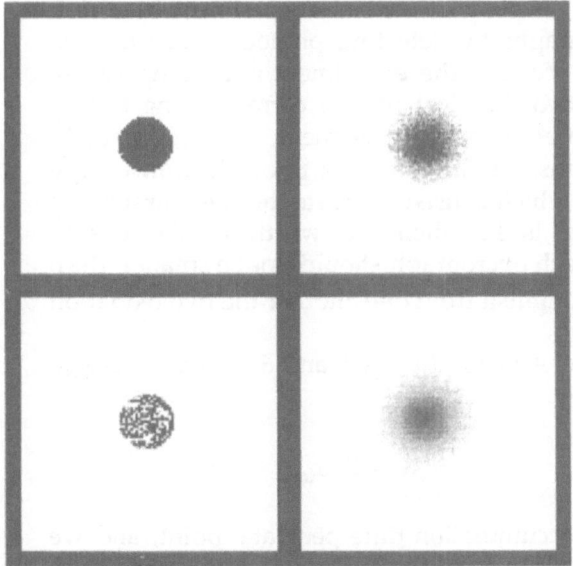

Figure 3. Optimization of the maximum (extractable) information content of a two-dimensional image: the "real object" is at the top left; the optimized information image is at the top right; the bottom images correspond to an overkill choice of space resolution, causing high noise level (bottom left) and to insufficient space resolution (bottom right).

Suppose, for example, that one takes the image of a dot with intrinsic width σ_R, using an "overkill" instrumental resolution $I/10$. This increases the entropy of Eq. 6 approximately by $(k/2)\ln(5)$ with respect to the optimum value. In order to offset this increase, one would have to increase the magnitude of the A-factor, and therefore the accumulation time per data point τ, by a factor of 5. The increase of the total data-taking time would be by a factor of 500 for a 100-pixel line-image. For a 100×100 pixel micrograph, it would by a potentially disastrous factor of 50,000: the mistake of overkilling the resolution problem could make it impossible to conduct the experiment!

The transition from microscopy to spectromicroscopy can be accomplished by using the same type of analysis. The analysis changes from one specific spectromicroscopy experiment to another; for example, from taking spectra in a number of microscopic areas to taking a micrograph by detecting photoelectrons of a given energy. The general message, however, is always the same: missing the optimum experimental conditions unnecessarily increases the data taking time, and the increase for spectromicroscopy is often higher than for two-dimensional microscopy.

Consider, for example, a spectromicroscopy experiment in which one takes two micrographs by detecting photoelectrons at two different energies; the scope of the experiment is to compare the space distribution of two chemical phases corresponding to two different oxidation states of the same element. The difference in the oxidation state causes a difference in photoelectron energy for a given core level, which is used to create the micrographs. If this energy difference is Δx, then the width of the energy window σ_{Ie} used for each micrograph should not be smaller than Δx, as required to distinguish from one another the two oxidation states.

The reason is that the A-factor in Eqs. 5 and 6 increases on σ_{Ie}, quite often linearly:

$$A \propto B\tau\sigma_{Ie} , \qquad (7)$$

where τ is again the accumulation time per data point, and we explicitly considered the role of the photon source's brightness, B. If one "overkills" σ_{Ie} by a factor of 10, this mistake must be compensated by an increase in the overall data taking time by the same factor; for $N \times N$ pixel micrographs, this goes like $N^2\tau$, and can easily lead to disaster.

5. Conclusions: the Role of Brightness

A complete analysis of the data taking strategy optimization for all possible spectromicroscopy experiments would be quite lengthy and, to some extent, not realistic, since subtle changes from one experiment to another can dramatically affect the conclusions. A series of cases are analyzed in Ref. 2, and I believe that the following general strategy conclusions can be derived from them:

- It is crucial to clear define the objective(s) of each spectromicroscopy experiment, and specifically the information that one wants to obtain.

- Then, one must select the energy and space resolution parameters trying to optimize the information content of the results. A good rule of thumb is to use an energy resolution comparable to the distance in energy of the spectral features that one wants to distinguish, and a space resolution comparable to the distance in space of the geometric features that one wants to recognize.

- Then, one has to decide how many spectra and what microimages are necessary to obtain the information which is the objective of the experiment.

- If several types of information are desired, one must optimize the data taking strategy for each type, without mixing up one strategy with another. In fact, the optimization for one type of information could lead to a strategy very far from optimization for another type, and therefore to a waste of time.

- In each case, the data taking strategy must be based on educated common sense, and never on brute force such as using "overkill" resolution levels.

What is the role of high brightness synchrotron sources in all this? This point can be analyzed by considering again the specific case of Eq. 6; as we have found in Eq. 7, the A-factor that links the signal level and the resolution also depends on the source's brightness. With high brightness, the signal-to-noise ratio increases and so does the information content.

In practice, this means that experiments that would be impossible with low or medium brightness sources become

52

possible with high-brightness sources like ELETTRA. Consider again Eq. 7: one could increase the magnitude of A, and therefore decrease the entropy and increase the information content, by increasing the accumulation time per data point, τ. This, however, could lead to an impossibly long data taking time for the entire experiment. Alternatively, one can increase the brightness of the photon source, keeping the data taking time within reasonable limits. In essence, a "reasonable" data taking time, which is often good but not strictly required for spectroscopy, becomes an absolute necessity for spectromicroscopy.

On the other hand, one should not fall in the trap of using the high brightness to cover one's strategic mistakes. In fact, the beamtime of ultrabright sources is extremely valuable and should never be wasted. This is true for every experiment, but particularly so in spectromicroscopy because of the perverse enhancement mechanism that transforms a small waste of time per data point into a disastrously large waste of time for the entire experiment.

In summary, ultrabright photon sources like ELETTRA open up very exciting new research opportunities in this general domain of spectromicroscopy -- but they cannot eliminate the need for a carefully planned data taking strategy, along the lines discussed above.

Acknowledgments

The author's own research in spectromicroscopy is supported by the Fonds National Suisse de la Recherche Scientifique, by the US National Science Foundation, and by the Ecole Polytechnique Fédérale de Lausanne.

References

1.. Margaritondo, G. (1994) Synchrotron Radiation and Free-Electron Laser Surface and Interface Spectroscopy and Spectromicroscopy, *Progr. Surf. Sci.* **46**, 275-293, and the references therein.
2. Margaritondo, G. (1995) The Information Content of Photoemission Spectroscopy and Spectromicroscopy, *Surface Rev. Lett.* (in press).
3. Gozzo, F. , Berger, H. , Collins,I. R., Margaritondo, G. , Ng, W., Ray-Chaudhuri, A. K., Liang, S., Singh, S., and Cerrina, F. (1995) Microscopic-Scale Lateral Inhomogeneities of the GaSe-Ge Heterojunction Energy Lineup", *Phys. Rev. B* (in press).

SCANNING SPECTRO-MICROSCOPY WITH 250 TO 800 eV X-RAYS

H. ADE,
*Dept. of Physics, North Carolina State University,
Raleigh, NC 27695*

C.-H. KO
*Dept. of Physics, SUNY@Stony Brook
Stony Brook, NY 11794*

Abstract

We describe spectromicroscopy experiments at the X1A facility at the National Synchrotron Light Source. In these efforts, two scanning x-ray microscopes based on zone plate optics are utilized. One of these is an ultra-high vacuum scanning photoemission microscope (SPEM) solely intended for surface characterization, while the other is a scanning transmission x-ray microscope (STXM). The former was substantially modified during the last few years and recently completed the first tests. We will describe its performance and outline further improvements. STXM has been utilized to provide chemical contrast via absorption spectroscopy in studies of a variety of polymeric materials in transmission. A spatial resolution of 30-40 nm and an energy resolution of 400 meV could be achieved in these experiments.

1. Introduction

1.1. GENERAL BACKGROUND

X-ray photoemission and x-ray absorption near edge spectroscopy (XPS and XANES, respectively) have evolved into powerful surface analysis techniques during the last two decades that provide information about the electronic and geometric structure of the sample investigated. A multitude of techniques and detection modes are presently practiced, which range from resonant photoemission spectroscopy to absorption spectroscopy via total electron-, Auger electron-, and fluorescent yield detection. Several efforts are now underway worldwide to achieve these spectroscopic capabilities at high spatial resolution [1-5]. One of the strategies employed is the utilization of zone plates to provide a finely focused probe of x-rays whose dimensions determine the spatial resolution. A mechanical raster scan will then provide

R. Rosei (ed.),
Chemical, Structural and Electronic Analysis of Heterogeneous Surfaces on Nanometer Scale, 53–74.
© *1997 Kluwer Academic Publishers.*

two dimensional information and the traditional detection schemes are utilized to provide spectroscopic information. Zone plates are also the basis of the spectro-microscopy efforts at beamline X1A at the National Synchrotron Light Source (NSLS). Major advantages of present high resolution zone plates are that they work well in the 200-1200 eV energy range, can be tuned over this energy range, and have already achieved a spatial resolution of 30 nm in transmission [6]. This paper will focus on the X1A spectromicroscopy efforts, particularly how they relate to the characterization of surfaces.

The motivation for photoemission microscopy of surfaces may be summarized as follows:

a) There is currently no other routine surface analysis tool that possesses submicron spatial resolution and is capable of providing elemental and **chemical** information from materials which are sensitive to charging and radiation damage, such as ceramics, polymers, and other insulators.

b) Even on conducting samples, XPS and XANES in many cases provide valuable information (detailed electronic structure of occupied and unoccupied states as well as bond orientation) which is not available with other surface microscopies, such as Auger microscopy, or Secondary Ion Mass Spectroscopy (SIMS) imaging, or even Scanning Tunneling Microscopy (STM).

Since zone plate based photoemission microscopes are scanning instruments, the detection efficiency of XPS and XANES detection modes can (in principle) be maximized and does not depend on the spatial resolution. In addition, fluorescent and luminescent photons could be detected. Charging and non-planar samples and surfaces are also relatively minor problems in scanning instruments, because they will at most affect the spectrum or the detection efficiency, but will not degrade the spatial resolution. Considering energy tunability, present and anticipated spatial resolution, sample constraints and other important parameters defining the various spectro-microscopy approaches, zone plate based photoemission microscopes seem to have one of the smallest numbers of physical limitations and should therefore result in one of the most general purpose photoemission microscope. Complementary photoemission microscopes on the other hand might be more optimized for specific tasks, and one should therefore pursue the development of these complementary approaches in parallel. A corollary is also true with respect to other complementary microscopies, which, like STM and AFM, might have better spatial resolution, but generally lack the chemical sensitivity provided by photoemission techniques. All these techniques should be developed and should compete with each other.

We will review in the following the present status of x-ray spectro-microscopy at beamline X1A of the National Synchrotron Light Source, outline the technological challenges involved and discuss present and potential future applications. We will exclusively focus on these efforts since many other approaches to x-ray spectro-microscopy are represented and discussed in these proceedings [5].

1.2. OVERVIEW OF X1A SPECTROMICROSCOPY PROGRAM

The X1A facilities at the NSLS are dedicated to soft x-ray imaging experiments [7, 8] including a wide range of experimental techniques. These make use of the NSLS soft x-ray undulator which is a high brightness source of soft x-rays. Its brightness has, however, been recently exceeded by third generation synchrotron radiation facilities. The first x-ray spectromicroscopy experiments at X1 were performed with the X1-SPEM-I, the first generation scanning photoemission microscope at X1A. This system was the first ultra-high vacuum (UHV) photoelectron microscope capable of chemical sensitivity at submicron resolution [9]. The resolution was subsequently improved to about 0.15 microns, first in one dimension [10], and later in two dimensions (astigmatism in the beamline optics was removed using elliptical zone plates) [11]. Although it has worked well for proof-of-principle experiments, this device had poor energy resolution (>4 eV), as well as poor acceptance.

Since it was apparent that substantial improvements in the performance of the X1-SPEM-I could be achieved with the incorporation of a commercial hemispherical sector analyzer (HSA) with multi-channel detection [10, 12, 13] only a limited number of experiments have been performed with this instrument. These experiments included, for example, the observation of photon-beam induced modifications of oxygen pre-dosed Al surfaces [13]. During the past years, Ko et al. [14, 15] constructed and commissioned the second generation SPEM (X1-SPEM-II) of which the first phase, which included the installation of a vastly improved spectrometer and more flexible zone plate mounting hardware, was recently completed. In parallel to the X1-SPEM-I upgrade, x-ray spectromicroscopy efforts at the X1A facility rapidly evolved around XANES microscopy in transmission of thin sections of primarily polymeric materials [16-21]. These experiments were performed with the Scanning Transmission X-ray Microscope (STXM) [8] at X1A in an atmospheric pressure environment. An energy resolution of 400 meV and a spatial resolution of about 30-40 nm has been achieved [6, 16, 18, 22]. We believe this to be the highest spatial resolution x-ray spectromicroscopy results. Although, these results are not obtained on surfaces, the same contrast mechanism can also be employed on surfaces and we will therefore discuss some representative STXM experiments after the presentation of the photoemission microscopy efforts.

2. Scanning photoemission microscopy at X1A

We have elected to build a scanning photoelectron microscope over an electro-static imaging x-ray microscope primarily because the scanning geometry provides the option of detailed analyses of pre-selected small spots on the surface (photo-electron spectra, absorption spectra, photon stimulated desorption studies) while not exposing the uninvestigated areas to the potentially damaging x-ray beam. In addition, scanning instruments permit acquisition of images with photoelectrons of preselected energy, by exploiting XANES in fluorescent-, total electron-, and Auger electron yield, and by monitoring the sample luminescence. For thin samples the transmitted flux can also be utilized. (Kunz's group at Hasylab/DESY in Germany has so far been most successful in implementing all the various detection modes of a scanning x-ray microscope [3, 23].) The detection of uncharged particles, such as fluorescent and luminescent photons is impossible with electro-static or magnetic imaging devices. In scanning devices, the spatial resolution and detection efficiency in any operating mode are furthermore not dependent on each other, and the detection efficiency tends to be large (which is of importance for radiation sensitive materials). Scanning x-ray microscopes, as compared to electro-static imaging microscopes, can also be used to investigate corrugated or insulating surfaces, consisting, for example, of etched trenches and other features of microelectronic devices. While our instrument will be able to operate in modes other than photoemission (such as transmission, and eventually fluorescent and luminescence detection), we will, for simplicity, continue to refer to it as the X1-SPEM-II.

2.1. DESCRIPTION OF THE X1-SPEM-II

The basic operating principle of the X1-SPEM-II is centered on the ability to focus the monochromatized x-ray beam from the X1 undulator to a microprobe through the deployment of a high resolution zone plate (ZP) (see schematic in Fig. 1).

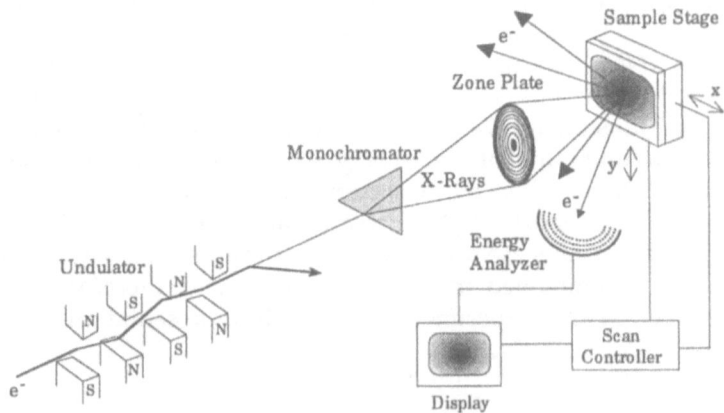

Fig. 1. Schematic representation of the X1-SPEM-II in XPS detection mode. Other possible but only partially implemented detection modes are the various XANES modes (Total electron yield, Auger electon yield, fluorescent yield), transmission, and luminescence spectroscopy. (Note that the energy analyzer is a hemispherical sector analyzer).

Photoelectrons emitted from the surface are then analyzed by a Hemispherical Sector Analyzer (HSA). XPS spectra from the small illuminated region can thus be obtained. With the HSA set to monitor a particular spectral range and feature, the specimen may be scanned to form images. Alternatively, the photon energy might be tuned to chemically specific absorption resonances to provide chemical maps, or the photon energy might be scanned to acquire an absorption spectrum from a small area.

All components of the X1-SPEM-II instrument reside inside a UHV chamber or are attached to it via ports and feedthroughs [24]. High throughput XPS capabilities are achieved with a Perkin Elmer HSA (model 10-360), equipped with an OMNI-II small area lens modified for high transmission and a multichannel detector (MCD, 16 channels). Since the photoelectrons are energy dispersed in the exit plane of the HSA, the use of a MCD allows us to monitor photoelectrons of different kinetic energies simultaneously. A total of 16 XPS images could thus be acquired in a single mechanical raster scan (see Fig. 2). The energy spacing of these images can be adjusted via the analyzer pass energy and can easily be matched to the photon energy resolution. One can thus monitor chemical shifts of a selected element and their intensities without any registration problems between images. The X1-SPEM-II presently utilizes a 16 channel Multichannel Scalar (MCS), to monitor up to 16 input signals simultaneously in order to form images.

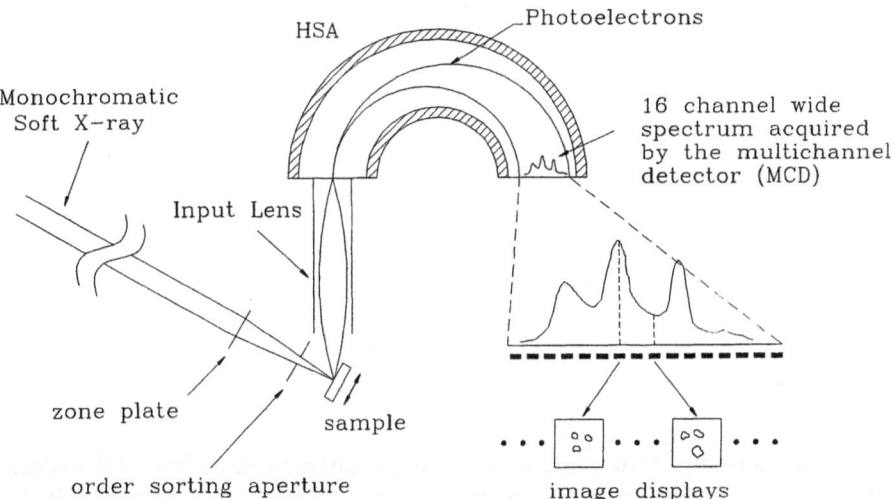

Fig. 2. Schematic of parallel photoelectron imaging utilizing an hemispherical sector analyzer with a multichannel detector.

Besides most of the XPS intensities, we typically monitor the sample replacement current, and the dwell time per pixel (clock), as well as the sum of all the 16 channels of the HSA detector, which is provided by the hemisphere electronics as a separate output. Hence, most channels of the MCD, i.e. ch. 3 to ch. 15, are connected to the MCS input signal channels for the parallel acquisition of 13 photoelectron images at different kinetic energies. An additional 3 input channels would be required to utilize all 16 detector channels of the MCD, while still monitoring the sample current, the clock, and the HSA sum.

One of the principle short-comings of zone plates for use in photoemission microscopes is their short focal length and therefore working distance (see Fig. 3). In addition, the focal length of the zone plate is proportional to the photon energy, and the sample distance has to be adjusted during acquisition of an absorption spectrum or any other photon energy change for the sample to stay in focus. With the present zone plates employed, the working distance is about 1-2 mm depending on the photon energy utilized. The mounts and manipulators for the optical components were carefully designed [24] to allow operation over the full energy range of a monochromator (250-800 eV) presently under construction. (The present "branchline" at which the X1-SPEM-II is installed is essentially limited to an energy range of 450-800 eV and the optical mounting mechanism has not been tested over the full design range). With some modifications to the zone plate mounting hardware, adjustments will be rapid and accurate enough to perform x-ray absorption spectroscopy from small areas, and to rapidly change energy over the full energy range. Useful auxiliary equipment such as an Ar-ion sputter gun, an electron gun, a transmitted flux detector, a sample current monitor, sample heating, electrical feedthroughs, gauges etc. is also available.

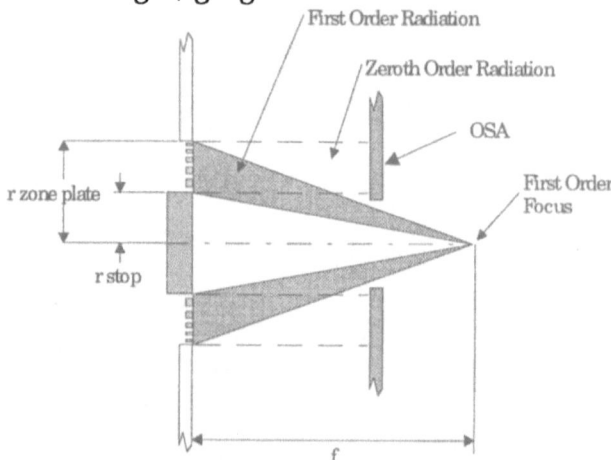

Figure 3. Side view of a zone plate and two of its diffraction orders. All orders other than the first order are blocked by the Order Selecting Aperture (OSA) placed in the shadow of a central zone plate stop. The first order focal spot is utilized as the microprobe in our instrument. A typical focal length of presently utilized zone plates is about 5-7 mm, resulting in a working distance of about 1-2 mm.

With a soon-to-be implemented monochromator at X1A, the photon energy of the X1-SPEM-II will be able to be tuned independently from other X1 branchlines between 250 and 800 eV. This energy range covers the carbon, nitrogen and oxygen K absorption edges, which will make it possible to perform XANES spectroscopy of these important elements with our instrument. In fact, care was put into the second generation SPEM design to eventually allow for the powerful combination of XPS and XANES analysis. The new monochromator will be optimized for imaging experiments, so that energy resolution can be traded for flux without affecting the spatial resolution. The spectral resolving power $E/\Delta E$ will be controlled solely by the entrance slit, and can be adjusted from about 500 to 4000. This feature results in a loss (or gain) of photon flux in the zone plate focal spot that is essentially linear with energy resolution. In contrast to operation at the soft x-ray undulator at the NSLS, which is not spatially coherent, this trade-off between energy resolution and available photon flux is not easily achievable at undulator at third generation sources because these sources are almost spatially coherent and the entrance slit can not be overfilled without penalty.

2.2. INITIAL PERFORMANCE OF THE X1-SPEM-II.

First results with the X1-SPEM-II validate our expectations of instrument performance. The effective data rate has increased by at least an order of magnitude over the first generation instrument, while simultaneously improving the signal-to-background ratio and the energy resolution [24]. We will show below some of these early results acquired from an integrated circuit of Nb on Si [15], and lines of Al, Si, and SiO_2 on Si [14]. For rapid progress during commissioning, the monochromator slits were opened up for these tests which resulted in worse than necessary energy resolution of about 1.8-2.8 eV for the selected incident photon energy of 540-689 eV. In many instances spectra were acquired in a matter of minutes, and improved energy resolution in actual experiments could be achieved with reasonable data acquisition times.

As described in Sec. 2.1, sixteen simultaneous images are typically acquired which was also the case of the Nb circuit (see Fig. 4). The energy separation between adjacent MCD channels is selectable by adjusting the pass energy. For this data set the energy separation of the HSA images is adjusted to 0.8 eV. The labeled channel number associated with each image is the MCD channel number. The kinetic energy of ch. 8 is 336 eV, and the remaining channels are spaced at 0.8 eV increments. These images cover thus the Nb 3d photopeak from 332.4 eV to 341.6 eV in kinetic energy. While chemical-shifts or changes in elemental composition and their associated relative intensity changes are not readily discernible for this particular sample, careful inspection shows that there are relative contrast changes between the round features and the rest of the sample (they are relatively bright in chs. 3-6, and relatively dark in most other channels). This is a reflection of their different composition as will become apparent below. These images give an indication of the baseline performance and the achievable signal to noise with the X1-SPEM-II in each of the 13 photoemission images. This parallel imaging technique recently demonstrated by Ko et al. [14] should provide high efficiency to map chemical core level shifts on material surfaces of a selected

60

element, and avoids any potential problem due to drifts and distortions that arise in data sets acquired in separate mechanical scans. In addition, it will allow to differentiate topographical contrast from chemical contrast relatively easily by image processing of images above, below, and at the photoelectron peaks.

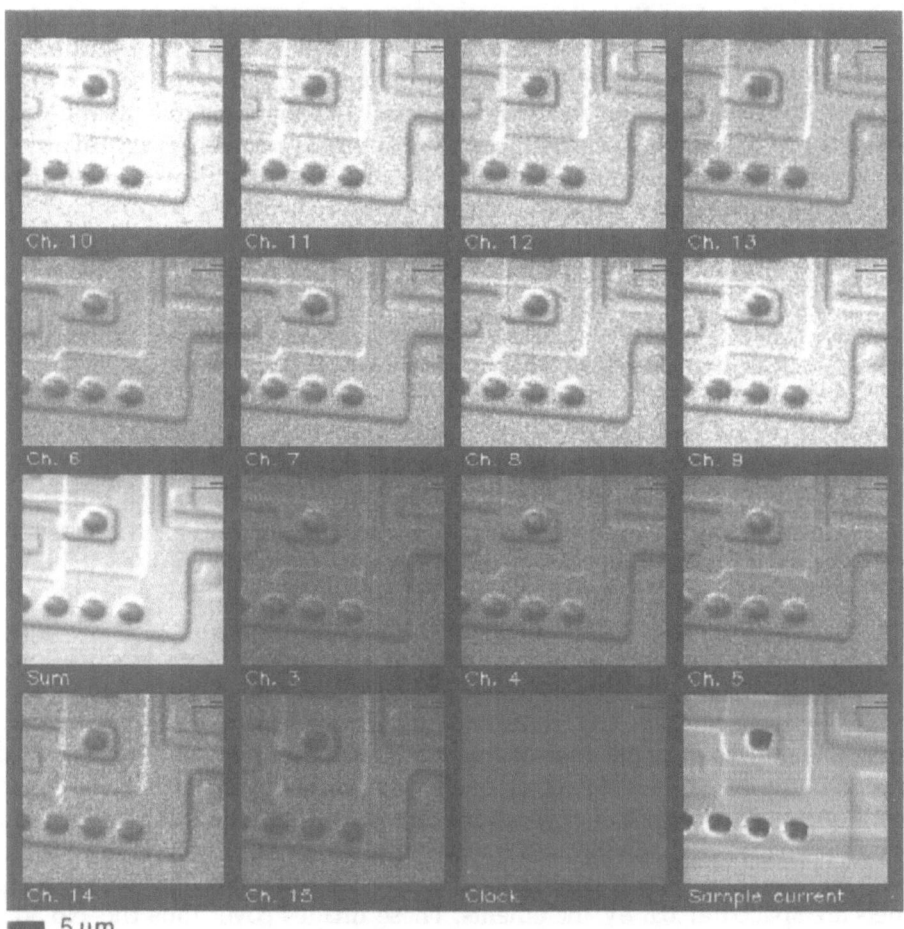

5 μm

Figure. 4. Images of Nb circuit on Si. The image labeled 'sum' is formed by monitoring an electronic output signal that carries the sum of all 16 MCD channels. Other images are labeled according to their MCD channel number.

Images were also acquired with the HSA set to monitor the Si-2p photoelectron peak. These images as well as XPS spectra from individual locations clearly show different elemental distributions in this sample [15]. To contrast these differences, we show the 'summed' Nb-4d image in Fig. 5a, and the 'summed' Si-2p photoelectron image in Fig. 5b. Many of the features observed are topological in nature. However, the round features in the sample

addition, the sample current and the pixel dwell time (clock) are monitored. (120x120 pixels x 0.25 μm, dwell time 220 msec/pixel, total acquisition time about 52 min.)

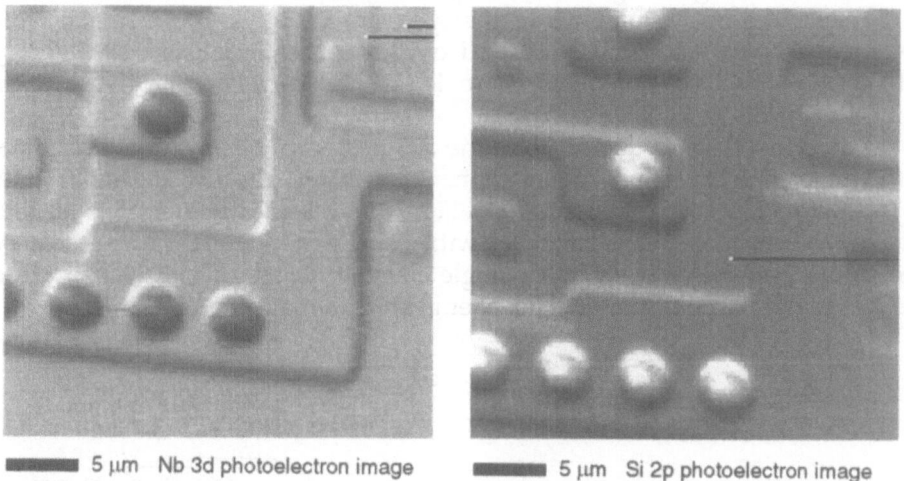

5 μm   Nb 3d photoelectron image   5 μm   Si 2p photoelectron image

Figure 5 a) 'Summed' photoelectron image utilizing the Nb 4p photoelectrons, b) 'summed' photoelectron image of about the same area as in Fig 5a with the HSA tuned to the Si 2p photopeak. (Same scan parameters as in Fig. 4.)

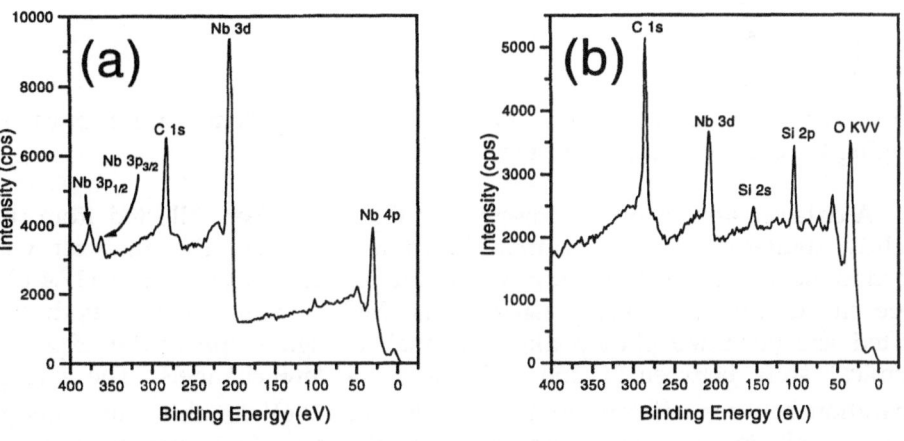

Figure 6. (a) spectrum from area outside the hole, (b) spectrum from inside one of the 3 μm diameter holes. The most prominent photoelectron and Auger peaks are identified and labeled. Note, that the O KVV Auger and the Nb 4p are overlapping, but are at slightly different energies. (Photon energy was 540 eV). The acquisition time for each spectrum was a few min (400 msec dwell time per 1 eV data point) and corresponds to a probe about 250 nm FWHM in size.

are holes that expose the Si substrate underneath and exhibit a quite different composition. Micrographs 5a and 5b clearly have areas with significant contrast reversal and the different composition is confirmed by the spectra shown in Fig. 6. Fig. 6a is the spectrum acquired from an area outside the 3 micron diameter holes while the spectrum in Fig. 4b is acquired inside one of these holes. The Si, C, Nb, and oxygen concentration is significantly different in these locations, and there is only appreciable Si concentration inside the hole.

Additional tests to demonstrate the acquisition of simultanous images of different chemical states of an element utilizing the MCD have been performed with the sample depicted in Fig. 7. As discused above up to 16 XPS images can be acquired simultanously with a single mechanical scan and can be used to monitor chemical shifts of single element. Ko et al. [14] have called this technique parallel-imaging for chemical state mapping (PICSM).

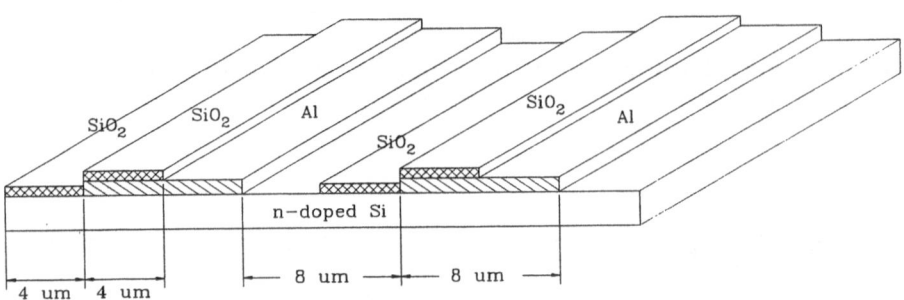

Figure 7. Schematic of the test pattern utilized to demonstrate parallel imaging for chemical state mapping.

A data set similar to that presented in Fig. 4 has been collected from the slightly sputter cleaned test sample shown in Fig. 7 [14]. The analyzer was tuned to the nominal Si 2p photopeak and the pass energy was set to 117.4 eV, large enough that the energy range of the MCD channels covers both the shifted and unshifted photopeaks, and small enough to spread the expected chemical shift between Si and SiO_2 across a few channels. The energy separation between adjacent MCD channels was 1.0 eV for this pass energy. Four special images were extracted from the set of 16 images and are displayed in Fig. 8. The scanned area covers slightly more than three lines, each of which is about 4 μm wide and can easily be observed in both the sample current image (Fig. 8a) and the 'summed' photoelectron image (Fig. 4b). Contrast reversal can be clearly observed between the photoelectron images of ch. 5 (Fig. 8c) and ch. 10 (Fig. 8d), demonstrating chemical state imaging. Detailed chemical analysis was obtained by acquiring XPS spectra from sub-micron areas in the locations as marked in Fig. 8 (i.e., location A is on the Al region, location B in the Si region, and locations C and D are in the two different SiO$_2$ regions). These spectra are displayed in Fig. 9, and were acquired with an incident photon energy of 689 eV, with an analyzer energy resolution of 2.0 eV and a photon energy width of 2.8 eV.

Al Si SiO$_2$ SiO$_2$

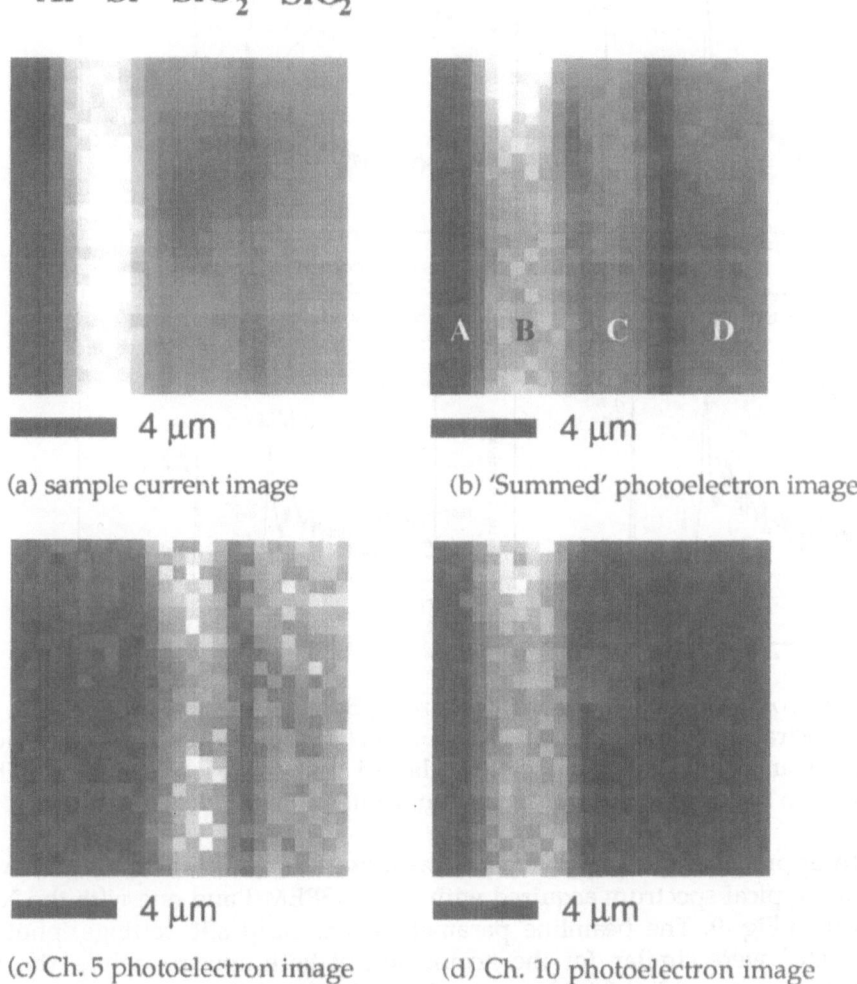

(a) sample current image   (b) 'Summed' photoelectron image

(c) Ch. 5 photoelectron image   (d) Ch. 10 photoelectron image

Figure 8. Images extracted from a data set similar to that shown in Fig. 4 demonstrating the mapping of different chemical states of the same element with a single mechanical scan. (a) sample current image, (b) 'summed' photoelectron image with an energy width of 16 eV, (c) photoelectron image of ch. 5, whose kinetic energy corresponds roughly to the oxide-shifted Si 2p, thus emphasizing the SiO$_2$ region, (d) photoelectron image of ch. 10, emphasizing the Si region, since its kinetic energy corresponds roughly to the unshifted Si 2p. Contrast reversal between the photoelectron image of ch. 5 and ch. 10 are clearly observed. (Image size is 20x20 pixels with a pixel size of 0.5 μm and dwell time of 300 msec per pixel, for a total acquisition time of about 2.5 min.)

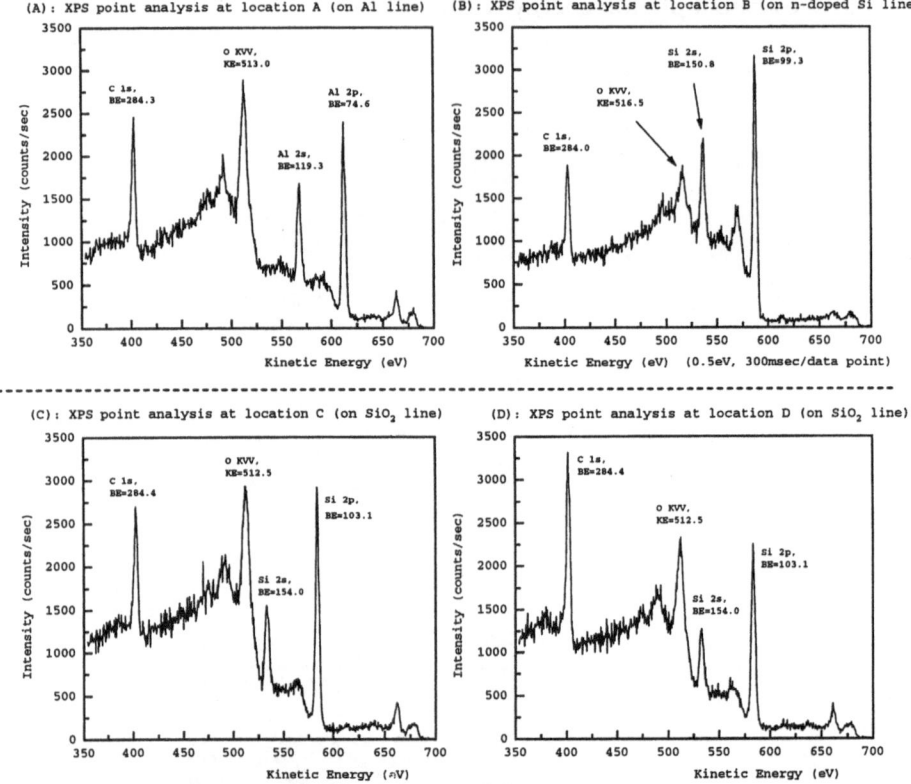

Figure 9. Point XPS analysis from locations A, B, C, and D as labled in Fig. 8. These spectra were acquired with 0.5 eV and 300 msec per data point, which results in an acquisition time of less then 4 min for each spectrum. The ordinates in the spectra are the actual count rates.

To appreciate the improvements made with the commercial HSA we present a typical spectrum acquired with the X1-SPEM-I and one with the X1-SPEM-II in Fig. 9. The beamline parameters (i.e. SGM slit settings, photon energy, etc.) were similar for the acquisition of both spectra, while the Si sample was more oxidized in the case of the X1-SPEM-II spectrum. Also, the zone plate has changed, but is similar in its main parameters. and we expect the zone plate efficiency to be about the same for both zone plates utilized. Whatever the reasons for improvements and taking the difference in sample into account, the comparison shows a combined improvement of over an order of magnitude in count rate, while simultaneously improving the energy resolution, as well as the signal to background ratio.

Note, that the energy resolution in these examples is dominated by the photon energy spread. With the new beamline, the loss in photon flux will be only linear with photon energy resolution and the "intensity-cost" for improved

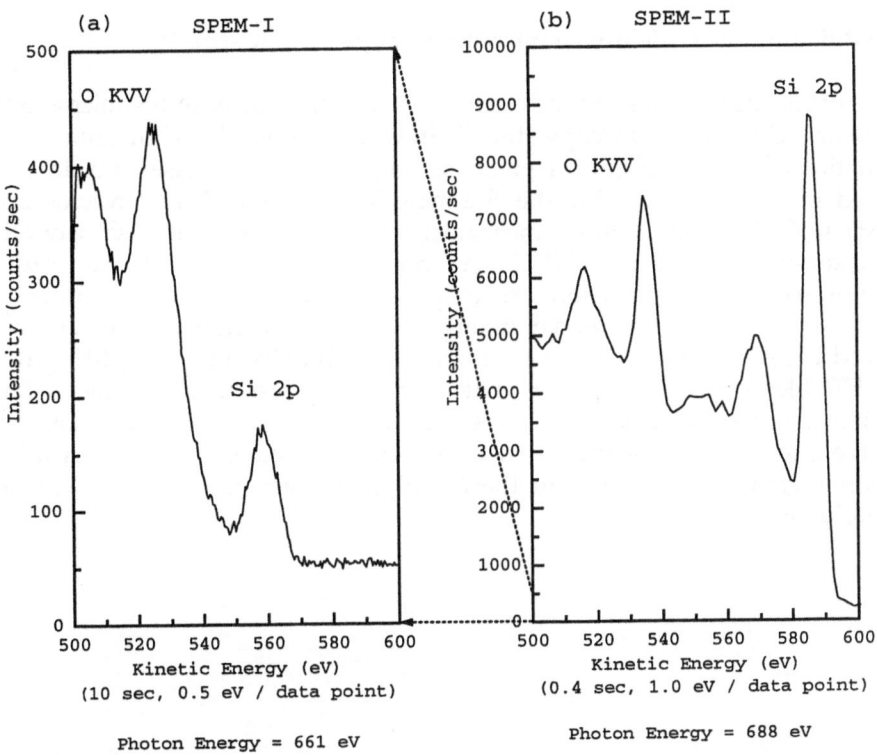

Figure. 10. Comparison of similar XPS spectra acquired under similar conditions with the X1-SPEM-I and X1-SPEM-II. While the spectra and acquisition conditions are not directly comparable, it is clear that the improvements with the X1-SPEM-II are more than an order of magnitude in photoelectron countrate, with a simultaneous improvement in energy resolution and signal to background ratio.

energy resolution will be quite tolerable. In addition, a choice of a lower photon energy would improve the energy spread in the photon beam by (energy ratio)^2 for the same monochromator slits. It will therefore be possible to achieve a combined energy resolution below 1 eV with sufficient intensity, since the available photon flux changes only with the grating efficiency, if the slits are kept at the same size. It is also important to note that due to a change in the beamline since the results in Ref. [11]and the lack of the right elliptical zone plates, X1-SPEM-II suffers from some astigmatism. With the

new beamline all these shortcomings will be removed and we should achieve
the diffraction limited resolution of the present zone plates utilized (about
100 nm) without any intensity losses in addition to the ones that come with
increased energy resolution. A somewhat more detailed description of this
situation can be found in reference [14].

3. XANES- and linear dichroism microscopy of polymeric samples

Other approaches to "spectromicroscopy" at X1A include XANES microscopy
and linear dichroism microscopy [16, 17], in which chemical and orientational
information of specific chemical groups is obtained. In both cases the data is
acquired in transmission with the Scanning Transmission X-ray Microscope
(STXM) [6,8], but the contrast mechanisms employed should also work if
surface sensitive modes of XANES microscopy via total/partial electron yield
and fluorescent yield detection are employed. An illustrative example of
XANES microscopy is presented in Fig. 11, where contrast reversal was
achieved between phases of a polycarbonate/ poly(ethylene terephthalate)
(PC/PET) blend by changing the photon energy by less than 350 meV. The
corresponding spectra of PC and PET as acquired with the STXM are shown in
Fig. 11, along with the chemical structure of the respective polymers. Similar
functional groups are present in these polymers, yet the spectra appear
strikingly different.

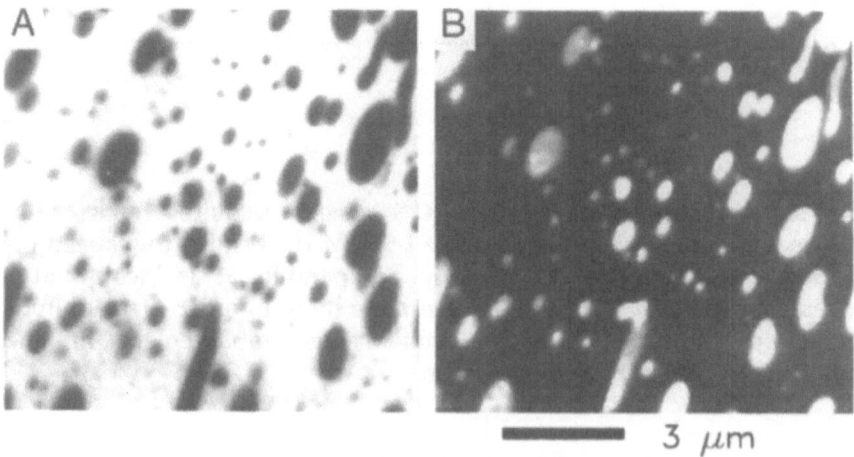

3 μm

Figure 11: Micrographs of PC/PET blend acquired at a photon energy of (A)
285.36 and (B) 285.69 eV. High contrast as well as contrast reversal with a
photon energy change of less than 350 eV has been achieved. (The STXM
monochromator was not fully calibrated at the time these images got
acquired) (Sample courtesy of G. Mitchell and B. Cieslinski, The Dow
Chemical Co.)

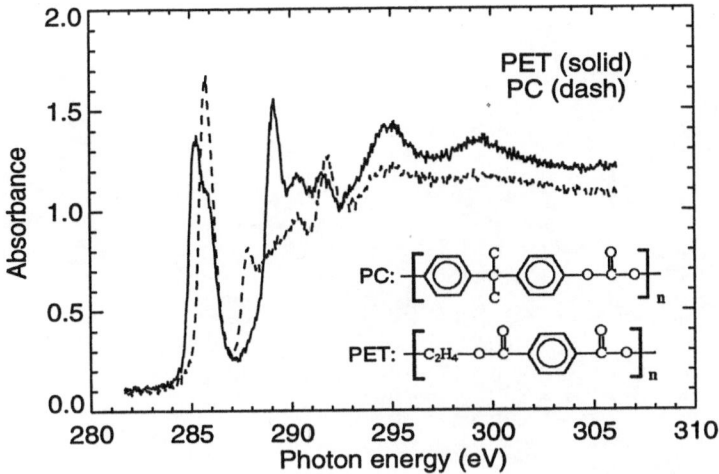

Figure 12: Spectra from 0.1 μm^2 areas of the 70/30 PC/PET blend shown in Fig. 11.. A rich set of spectral features characteristic of each polymer is observed. The spectra are quite different given the relative structural similarity of the respective polymers, and where not normalized for density or thickness variations of the thin film (PC: dash, PET: solid) [18]. (Note that these spectra have been acquired during the presence of at the time undetected monochromator non-linearity, which distorted and compressed the energy scale locally particularly between 286 and 305 eV, i.e. the actual peaks in this energy range are at lower energies).

As an additional example of XANES imaging of polymers, which are generally radiation sensitive materials, we present here the investigation of the phase morphology of a liquid crystalline polyester based on several aromatic monomers. Electron microscopy (EM) and differential scanning calorimetry indicated the existence of phase separation in materials based on this polyester, but these techniques could not determine whether the phases are chemically distinct, and if so, how many phases there are, or whether the observed phase separation is solely due to differences in crystallinity. We compared melt-screened and un-screened samples. In the melt-screening process, small particulates of a certain size (>25 μm) are removed from the polymer melt with a screen. We found that the un-screened material had four chemically distinct phases. As an illustration we show micrographs acquired at (Fig. 13a) 285.0 eV, (Fig. 13b) 286.8 eV, (Fig. 13c) 296.2 eV, and (Fig. 13d) 281.8 eV. Micrograph 13a emphasizes aromatic content and a discontinuous phase with domains smaller then 100 nm, as well as continuous phases with dimensions of a few microns are clearly discernible. Fig. 13b reverses the contrast between the continuous phases, while there is virtually no contrast between the small

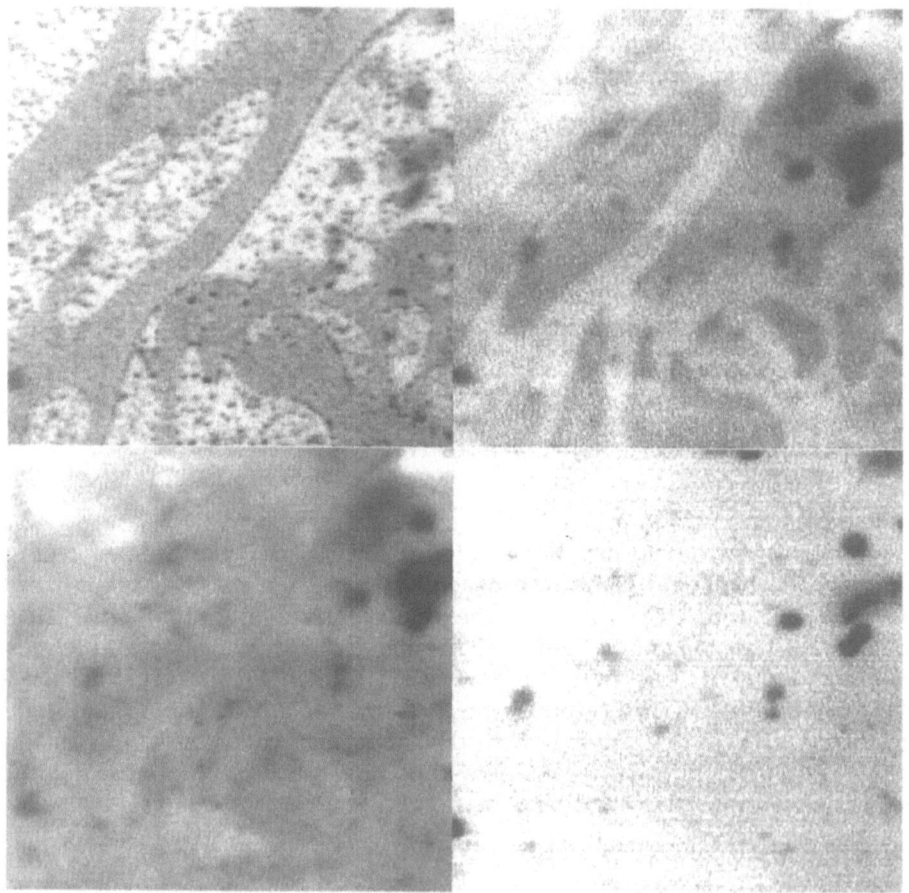

Fig. 13. Micrographs of the same region of a thin section of an aromatic liquid crystalline polyester imaged at a photon energy of (13a, top left) 285.0 eV, (13b, top right) 286.8 eV, (13c, bottom left) 296.2 eV, and (13d, bottom right) 281.8 eV. Field of view is 9 x 9 μm. (Sample courtesy of B. Wood, and I. Plotzker, DuPont)

features. Fig. 13c is acquired at a photon energy that has only residual chemical sensitivity and is predominantly a "density map" that most closely resembles the EM micrographs of this material. Micrograph 13d is acquired below the carbon edge, and emphasizes elements other than carbon. Given the elemental constituents and the low level of metal contamination in this polymer, the dark features in Fig. 13d are phases rich in oxygen. This completely unexpected and complex chemical morphology observed in the un-screened material was totally absent in the screened material, which illuminates and illustrates the structural dependence of the materials on the processing route. The observed results can not be explained by merely removing larger particulates from the melt, but the screening process itself must have altered the processing dynamics and in turn prevented phase separation.

In addition to the chemical characterization of various polymers and other materials [16, 18-21], linear dichroism microscopy has been successfully demonstrated [17] and utilized to reveal that the average aromatic ring plane in Kevlar fibers points radially outwards [17, 18]. This is exemplified by the butterfly pattern observed in micrographs of thin sections of Kevlar 149 fibers, presented in Fig. 14. Linear dichroism microscopy is the only analytical technique to determine orientational order quantitatively at high spatial resolution, and is presently being used to elucidate the differences in orientational order between the various grades of Kevlar fibers (Kevlar 149, Kevlar 49, and Kevlar 29). Quantitative information can be extracted most readily from spectra such as the ones shown in Fig. 14. Note, that the dichroism observed is much larger for Kevlar 149 than for Kevlar 49, and is near the detection limit for Kevlar 29 [25].

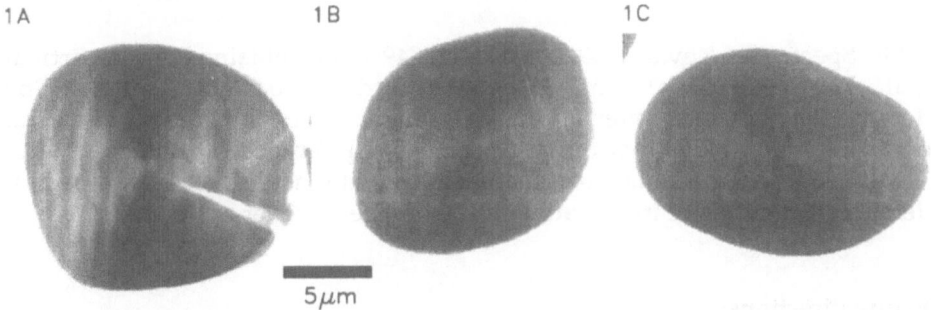

Figure 14. Micrographs of 200 nm thick sections (cut at 45° relative to fiber axis) of (A) Kevlar 149, (B) Kevlar 49, and (C) Kevlar 29, imaged at a photon energy of 285.5 eV. This energy is characteristic of the aromatic groups of the fiber polymer and the butterfly patterns observed in all three grades of Kevlar fibers are due to the radial symmetry and orientational order of the fibers. Images are slightly contrast enhanced and do not quantitatively represent the relative differences in orientational order. We extract quantitative information about these samples primarily from spectra as shown in Fig. 15. (Sample courtesy of B. Hsiao and S. Subramoney, DuPont)

While these XANES and linear dichroism images have been acquired in transmission in an atmospheric pressure environment (He enclosure), the same contrast mechanism could be employed in a vacuum set up to investigate surfaces. Total electron yield, fluorescence yield and Auger electron yield detection modes will be the principle avenues that can be explored, and if all three were installed would even allow one to vary the surface sensitivity.

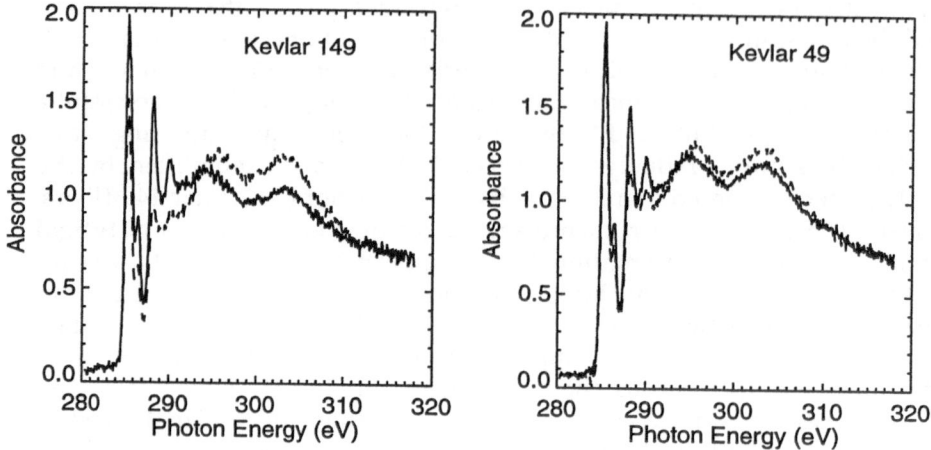

Fig 15: Spectra of Kevlar 149 and Kevlar 49 fibers obtained from vertical radial locations (dash) and horizontal radial location (solid). The differences in the peak intensities is an indication of the degree of radial order in the fiber. Note, that the peak intensity differences are much smaller in Kevlar 49 than in Kevlar 149. By fitting the intensities of these spectral differences, quantitative information of the degree of order can be obtained.

4. Future directions

4.1 OPTICS IMPROVEMENTS: X1A BEAMLINES AND ZONE PLATES

The undulator and the present beamline with two branch lines sharing one grating were built six years ago. The beamline itself is presently undergoing considerable redesign. The installation of two new beamlines, each with its own grating, is scheduled to take place during the 1995 end-of-the-year shutdown of the NSLS x-ray ring. Two fully independent beamlines will then be available to deliver x-rays simultaneously to two endstations, only sharing the gap setting of the undulator. Each independent beamline will share control of the gap on a time split basis. The relatively low brightness of the NSLS works actually in our favor in this regard. Not only is it possible to feed a total of three beamlines simultaneously (two at X1A and one at X1B) from the same undulator, but the undulator spectral output has considerable and usable flux for any given gap setting over about 30-40% of the energy range covered by the monochromators. With some planning and careful scheduling it should be possible to find agreeable operating parameters and undulator gaps that will satisfy the demands of the users on both X1A monochromators for most of the time. Careful design of the beamline to be utilized by the X1-SPEM-II will allow to eliminate astigmatism and the need to utilize elliptical zone plates. The beamline will also provide slightly higher photon fluxes (about 40%) for the same energy resolution (assuming the same grating efficiency) since the energy resolving power will be dominated by the entrance slit width with only a minor contribution by the exit slit. We expect the spatial resolution to be the diffraction limit of the respective zone plate utilized with this new beamline. We also anticipate some improvement in zone plate technology and our ability to master the short working distances of these devices. As pointed out previously, zone plates in transmission have already achieved a spatial resolution much superior to the X1-SPEM-II. As we gain experience with our new instrument, we anticipate to be able to take advantage of higher resolution zone plates and the spatial resolution to improve towards 50 nm.

4.2. OTHER IMPROVEMENTS OF THE X1-SPEM-II

There are still a variety of mechanical improvements possible to make the X1-SPEM-II a more versatile micro-analysis instrument. In particular, improvements to the zone plate mounting hardware would allow us to also acquire XANES spectra from small areas on the sample comparable to the spatial resolution. Presently, drifts and run-out significantly worsen the spatial resolution in XANES spectral mode. In addition, we anticipate an improvement of about a factor of two in XPS data rates by shortening the sample distance to the HSA and refocusing the HSA input lens, thus providing increased magnification and a larger solid angle collection. Raytracing of the input lens performed by Perkin Elmer indicate that we should gain almost a factor of three by utilizing this method. Effective charge neutralization for insulating and poorly conducting samples would also greatly increase the classes of samples that could be investigated.

4.3. POSSIBLE APPLICATIONS OF THE X1-SPEM-II

While Ade et al. [16-18] as well as others [19-21] have already successfully employed the spectro-microscopy capabilities of the STXM, the scientific program with the X1-SPEM-II is just beginning. The studies we envision to undertake with the X1-SPEM-II in collaboration with several colleagues are the physical and chemical characterization of micro- and nano-structured materials and their surfaces, as well as the dependence of the structure-property relationship of these materials on processing. We are particularly interested in characterizing: a) the surface properties, such as adhesion and permeability, of heterophase polymers, b) polymer-glass composites and their failure modes at the interfaces, c) adhesion failure and corrosion of metal/polymer interfaces and processing parameters of photo-imageable polymers in device packaging, d) worn and rubbed surfaces produced in the presence of various lubricant additives, and e) composition of plasma etched features in Si wafers, and how the determined properties depend on processing. The common majority components in these systems intended for initial investigations are carbon, nitrogen and oxygen, elements which exhibit a rich set of chemical and structural signatures when investigated with XPS and XANES techniques. Most of these materials and samples are also prone to charging and have corrugated surfaces. For these reasons, and the high photon energy required to assess the chemistry of the elements of interest. we think that these systems would be more difficult (if not impossible) to be analyzed with other implementations of x-ray spectromicroscopy and are therefore well matched to the capabilities of the X1-SPEM-II. In most cases, we also expect to require only a modest energy resolution.

5. Acknowledgments

We are grateful to the many people who have made the presented work possible. In particular we would like to thank J. Kirz, C. Jacobsen, X. Zhang, S. Wirick and their coworkers at SUNY, Stony Brook, who built and maintain the STXM, and who have greatly contributed to the implementation of the XANES operating mode of the STXM. We would also like to thank M. Rivers, S. Hulbert, E. Johnson, K. Maier, and B. Winn for their contributions to the X1-SPEM-I and/or the X1-SPEM-II. The zone plates for these experiments have been provided by E. Anderson and D. Attwood from the Center for X-ray Optics. We would also like to acknowledge and thank J.C. Lin, G. Mitchell, B. Cieslinski, B. Hsiao, S. Subramoney, B. Wood, and I. Plotzker for providing some of the samples shown, and A.P. Smith for help with data acquisition and analysis. This work is supported in part by the National Science Foundation under grants DMR-9458060 and DIR-9005893, by The Dow Chemical Corp. and EXXON Research and Engineering and by a DuPont Young Professor Grant. Work at the NSLS is supported by the US Department of Energy under contract DE-AC-02-76CH00016. The center of X-ray Optics is supported by the US Department of Energy under contract DE-AC-03-76SF00098.

References

1. For a somewhat dated overview that however nevertheless lists most of the spectromicroscopy efforts presently under way, see Ade, H.W., (1992) Scanning photoemission microscopy with synchrotron radiation, *Nucl. Instr. Meths. in Phys. Res. A.* **319**, 311. Also see reference [5] and articles and references therein for some of the newer developments.
2. Tonner, B.P., Harp, G.R., Koranda, S.F., and Zhang, J., (1992) An electrostatic microscope for synchrotron radiation X-ray absorption microspectroscopy, *Rev. Sci. Instr.* **63**, 564.
3. Kunz, C. and Voss, J., (1995) Scientific progress and improvement of optics in the VUV range, *Rev. Sci. Instrum.* **66**, 2021.
4. Ng, W., Ray-Chaudhuri, A.K., Liang, S., Singh, S., Solak, H., Welnak, J., Cerrina, F., Margaritondo, G., Underwood, J.H., Kortright, J.B., and

Perera, R.C.C., (1994) High Resolution Spectromicroscopy with MAXIMUM: Photoemission Reaches the 1000 Å Scale, *Nucl. Instrum. Meth.* **A347**, 422.

5. Rossei, R., (1995) *Chemical, Structural and Electronic Analysis of Heterogenous Sufaces on Nanometer Scale*, Klewer Academic Puplishers, Dordrecht. These proceedings.0

6. Jacobsen, C., Williams, S., Anderson, E., Brown, M.T., Buckley, C.J., Kern, D., Kirz, J., Rivers, M., and Zhang, X., (1991) Diffraction-limited imaging in a scanning transmission X-ray microscope, *Opt. Commun.* **86**, 351.

7. Rarback, H., Buckley, C., Ade, H., *et al.*, (1990) Coherent radiation for X-ray imaging-the soft X-ray Undulator and the X1a Beamline at the NSLS, *J. X-ray Sci. Technol.* **2**, 273.

8. Kirz, J., Ade, H., Howells, M., Jacobsen, C., Ko, K.-H., Lindaas, S., McNulty, I., Sayre, D., Williams, S., and Zhang, X., (1992) Soft X-ray microscopy with coherent X rays, *Rev. Sci. Instrum.* **63**,

9. Ade, H., Kirz, J., Hulbert, S., Johnson, E., Anderson, E., and Kern, D., (1990) X-ray spectromicroscopy with a zone plate generated microprobe, *Appl. Phys. Lett.* **56**, 1841-1843.

10. Ade, H., Kirz, J., Hulbert, S., Johnson, E., Anderson, E., and Kern, D., (1991) Images of a microelectronic device with the X1-SPEM, a first generation scanning photoemission microscope at the National Synchrotron Light Source, *J. Vac. Sci. Technol.* **9**, 1902.

11. Ade, H., Ko, C.H., and Anderson, E., (1992) Astigmatism correction in X-ray scanning photoemission microscope with use of elliptical zone plate, *Appl. Phys. Lett.* **60**, 1040.

12. Ade, H., (1990) Development of a scanning photoemission microscope, Ph. D thesis, SUNY@Stony Brook, Stony Brook.

13. Ade, H., Ko, C.-H., Johnson, E.D., and Anderson, E., (1992) Improved Images with the Scanning Photoelectron Microscope at the National Synchrotron Light Source, *Surface and Interface Analysis.* **19**, 17.

14. Ko, C.-H., Kirz, J., Maier, K., Winn, B., Ade, H., Hulbert, S., Johnson, E., and Anderson, E., (1995) Chemical State Mapping of Material Surfaces with the X1A Second Generation Scanning Photoemission Microscope (X1A SPEM-II), in W. Yun (eds.), *X-ray Microbeam Technology and Applications*, Proc. SPIE Vol 2516.

15. Ko, C.-H., Kirz, J., Ade, H., Johnson, E., Hulbert, S., and Anderson, E., (1995) Applications of the scanning photoemission microscope for element identification on material surfaces, *Mat. Res. Soc. Proc. Vol.* **375**, 303.

16. Ade, H., Zhang, X., Cameron, S., Costello, C., Kirz, J., and Williams, S., (1992) Chemical contrast in X-ray microscopy and spatially resolved XANES spectroscopy of organic specimens, *Science* **258**, 972.

17. Ade, H. and Hsiao, B., (1993) X-ray Linear Dichroism Microscopy, *Science* **262**, 1427.

18. Ade, H., Smith, A., Cameron, S., Cieslinski, R., Costello, C., Hsiao, B., Mitchell, G., and Rightor, E., (1995) X-Ray microscopy in polymer science: Prospects of a "new" imaging technique, *Polymer* **36**, 1843-1848.

74

19. Zhang, X., Balhorn, R., Jacobsen, C., Kirz, J., and Williams, S., (1994) Mapping DNA and Protein in biological samples using the scanning transmission x-ray microscope, in G.W. Bailey and A.J. Garratt-Reed (eds.), *Proc. 52nd Annual Meeting of Microscopy Society of America*, San Francisco Press, Inc., San Francisco, 50-51. Also, Zhang, X., Balhorn, R., Mazrimas, J., and Kirz, J. (1995) Mapping and measuring DNA to protein ratios in mammalian sperm head by XANES imaging, *Journal of Structural Biology*, (submitted).
20. Botto, R.E., Cody, G.D., Kirz, J., Ade, H., Behal, S., and Disko, M., (1994) Selective Chemical Mapping of Coal Microheterogeneity by Scanning Transmission X-ray Microscopy, *Energy & Fuels* **8**, 151-154.
21. Cody, G.D., Botto, R.E., Ade, H., Behal, S., Disko, M., and Wirick, S., (1995) C-NEXAFS Microanalysis and Scanning X-ray Microscopy of Microheterogeneitie in a High Volatile A Bituminous Coal, *Energy & Fuels* **9**, 153. Also,   Cody, G.D., Botto, R.E., Ade, H., Behal, S., Disko, M., and Wirick, S., (1995) Inner shell spectroscopy and imaging of a sub bituminous coal: in situ analysis of organic and inorganic microstructure using C(1s)-, Ca(2p), and Cl(2s) NEXAFS, *Energy & Fuels.* (accepted for publication)
22. Ade, H., Smith, A.P., Subramoney, S., and Hsiao, B., (in preparation)
23. Voss, J., Storjohan, I., Kunz, C., Woewes, A., Pretorius, M., Ranck, A., Sievers, H., Wedemeier, V., Wochnowski, M., and Zhang, H. (1994) Soft X-ray Microscopy at HASYLAB/DESY, in A.I. Erko and V.V. Aristov (eds.), *X-ray Microscopy IV*, Bogorodski Pechatnik, Chernogolovka, Moscow Region.
24. Ko, C.-H., (1995) Development of a Second Generation Scanning Photoemission Microscope at the National Synchrotron Light Source, Ph.D. Thesis, SUNY@Stony Brook, Stony Brook.
25. Smith, A.P., Ade, H., Subramoney, S., and Hsiao, B., (in preparation)

RECENT ADVANCES IN LEEM/PEEM FOR STRUCTURAL AND CHEMICAL ANALYSIS

E. BAUER, T. FRANZ, C. KOZIOL, G. LILIENKAMP AND
T. SCHMIDT
Physikalisches Institut, Technische Universität Clausthal
D 38678 Clausthal-Zellerfeld, Germany

Non-scanning surface electron microscopy with slow emitted and reflected electrons is slowly approaching the goals set more than five years ago. This paper discusses mainly the present state of art of imaging with a multimethod instrument which allows LEED, LEEM, MEM and various emission microscopies without and with energy filtering, using electrons or photons as primary species. It includes also a brief comparison with spin-polarized LEEM and with pure emission microscopy.

1. Introduction

It is now ten years that the first low energy electron microscopic (LEEM) images were published [1]. Although low energy electron diffraction (LEED) allowed in many cases chemical identification via material-characteristic LEED patterns, the desire soon arose to combine LEEM with a more chemically specific imaging mode such as Auger electron emission microscopy (AEEM) or core photo electron emission microscopy (CPEEM). It was obvious that good resolution and signal/noise ratio in AEEM and CPEEM required a band pass electron energy filter in the imaging column and that the illuminating beam in LEEM could also be used for AEEM if the beam separator would allow the passage of electrons with different energy in illuminating and imaging beam. W. Telieps´ untimely death in 1987 delayed the project until an experienced instrument designer was found in L.H. Veneklasen who solved the design requirements in an instrument reported in 1991 [2]. Technical problems delayed the test of the CPEEM mode at an undulator beam line until 1994.

In the meantime the first CPEEM results were already obtained in a conventional PEEM with straight optical axis and without energy filtering [3]. The poor lateral resolution due to the wide energy distribution of the photoexcited secondary electrons convinced us [4] that this mode of operation without energy filtering or correction of the chromatic aberration is not very useful because there are sufficient other, simpler methods with the same resolution. Unfiltered CPEEM will, therefore, only be briefly touched upon here, X-ray magnetic circular dichroism imaging (XMCDPEEM) excepted. This method, first reported in 1993 [5] gives information on the magnetization distribution on the surface (magnetic microstructure), similar to the kind of

R. Rosei (ed.),
Chemical, Structural and Electronic Analysis of Heterogeneous Surfaces on Nanometer Scale, 75–91.
© 1997 *Kluwer Academic Publishers.*

information which can be obtained from spin-polarized LEEM (SPLEEM) [6]. For this reason, SPLEEM will be discussed briefly too.

2. Instrumental aspects

The basic design considerations and their practical realization have been discussed extensively by Veneklasen [2,7] so that only the present configuration of the instrument as it is used on an undulator beam line will be described.

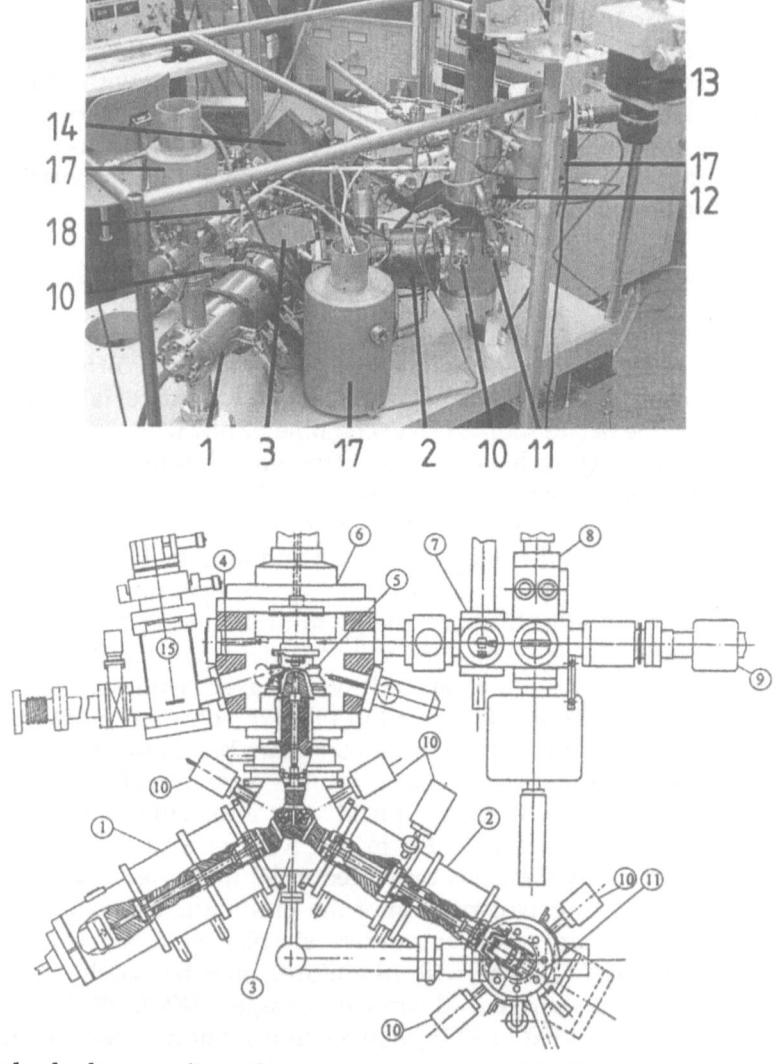

Fig. 1 Cathode lens surface electron microscope with beam separator and energy filter. a) Schematic, b) physical appearance. For explanation see text.

Fig. 1a shows the schematic of the instrument, Fig. 1b its physical set-up without the connections to the beam line. The angle between illumination column (1) and imaging column (2) is 120 °C. The beam separator (3) connects the two columns with the specimen chamber (4) which contains the cathode objective lens (5) and the specimen stage which is mounted on the specimen manipulator (6). The specimen is introduced into the specimen chamber through the specimen preparation chamber (7) via the airlock (8) using the transfer rod (9). Four aperture manipulators (10) operate (from left to right) the illumination aperture, the contrast aperture and the energy selection slit at the exit of the energy filter (11). The energy filtered final image is produced by the projective (12) on a double channel plate/fluorescent screen detector and recorded either by a newvicon video camera on tape (for high intensity, high contrast operation modes such as LEEM or LEED) or by a Peltier-cooled CCD camera (13) via a frame grabber on disc (for low intensity, low contrast operation modes such as AEEM or CPEEM). Photo electrons can be excited by two sources: valence electrons for VPEEM by a high pressure Hg short arc lamp (14) and core electrons by undulator radiation which is refocussed by a mirror in the mirror chamber (15).

The lens system is more complicated than in a standard LEEM system for two reasons. i) The illumination system has to be operated in two modes: a) a high coherence mode, in which the beam is as parallel and monochromatic as possible for LEED, LEEM and MEM and b) in a high intensity mode for AEEM, where coherence of the incident beam is unimportant but maximum intensity is needed. In order to achieve this flexibility three lenses are used: one to demagnify the cross-over of the LaB_6 electron gun, one to control the angular aperture and one to image, together with the double focussing beam separator, the demagnified cross-over either into the back focal plane of the objective (for LEED, LEEM and MEM) or, together with the objective lens, onto the specimen for AEEM and high intensity SEEM. The illumination aperture allows to reduce the illuminated area into the sub-micron range for selected area LEED so that no complementary field limiting aperture is necessary on the image side of the beam separator. On the imaging side several lenses are necessary so that the image can be placed with the optimum magnification and angular aperture into the proper position at the entrance of the energy filter. Four lenses are used to achieve this goal: a transfer lens images the LEED pattern, which is located in the back focal plane of the objective lens, into the center of a field lens where the contrast aperture is located. An intermediate and projective lens which act as a zoom lens then image either diffraction or image plane at the entrance of the energy filter which is a 90° retarding hemispherical analyzer. The contrast aperture was removed from the usual position in the back focal plane of the objective lens mainly not to limit the intensity in the AEEM mode but also to allow large off-axis displacements of the incident beam for tilted illumination of the specimen in LEEM which is useful for imaging with not specularly reflected LEED beams.

For AEEM primary energies E_p at the specimen up to 3 keV are needed if all elements are to be detected with optimum efficiency while the energies E_A of the characteristic Auger electrons can be as low as several 10 eV. Therefore, the beam separator must be able to deflect electrons with different energies in the illuminating beam ($E_1 = eU_o + E_p$) and in the imaging beam ($E_2 = eU_o + E_A$) by 60°. This is achieved by splitting the deflection field into two parts in form of a "close-packed prism array". Another unusual feature of the instrument is the objective lens which is a magnetic triode lens. In this lens the two magnetic pole pieces can be put at different electrostatic potentials. If the specimen can withstand a high electric field then both pole pieces are at ground potential and the full potential difference is applied between specimen and magnetic lens. This gives maximum resolution and collection efficiency. If the specimen cannot withstand very high fields the pole piece next to the specimen is put at a potential intermediate between specimen and ground potential. This gives considerable flexibility in the type of specimens which can be studied without changing the distance between specimen and magnetic lens. Keeping the specimen position fixed is important for many in situ experiments which make use of one of the six ports pointing at the specimen at 16° glancing angle. For example, the spot illuminated by the undulator or Hg light moves on the specimen when its axial position is changed. Similarly, evaporators or gas sources pointing at the specimen may produce coverage gradients or even shadows if the atomic or molecular beam is partially cut off in the wrong specimen position by the holes in the magnetic shielding surrounding the specimen.

The instrument is pumped with several sputter-ion pumps one of which (16) is shown in Fig. 1a and several Ti sublimation pumps, three of which (17) are visible in Fig. 1b. They produce base pressures in the 10^{-10} Torr range in the gun and detector chambers and in the 10^{-11} Torr range in the main chamber in the stand-alone instrument. When connected to the beam line, the pressure in the main chamber rises into the 10^{-10} range. The large Ti sublimation pump in the foreground of Fig. 1b pumps directly on the beam separator which is connected to the other pumps only by narrow tubes and is a major source of gas produced by electron stimulated desorption, in particular at the high beam currents used in AEEM. Thorough bake-out is needed if specimen contamination by back streaming desorption products is to be avoided, even if the pressure measured in the main chamber is in the low 10^{-10} Torr range. In order to reduce this problem, a small straight-through valve (18) is inserted between beam separator and main chamber which allows to keep beam separator, illumination and imaging column under UHV when the specimen chamber has to opened in order to change experimental components (evaporators, gas sources, etc).

More information on instrumental aspects may found in refs [2,7] and some recent reviews [8,9]. At present, there is only one instrument of the type described here which is equipped with an energy filter and two without a filter. The other operational LEEM instruments have simple beam separators for equal electron energy in illuminating and imaging beam. Three of them have the original 60° deflection design [1,10] while the most recent one uses a double 45° deflection [11]. This and the original LEEM instrument have been converted to SPLEEM with a GaAs photoemission electron source [6]. Several new instruments are presently in development.

3. Imaging without energy filter and split beam separator

This type of imaging is possible with all LEEM instruments. It includes LEEM, MEM, LEED, VPEEM, thermionic electron emission microscopy (TEEM) and emission microscopy with several other excitation modes, in particular secondary electron emission microscopy (SEEM) using the LEEM illumination beam for excitation. If the contrast aperture is located after the beam separator as described above, then it can be centered on the maximum of the secondary electron angular distribution for SEEM. By tilting the illuminating beam the maximum of the secondary electron angular distribution can be kept on the optical axis. This imaging mode has been used very little up to now but is an important alternative to VPEEM and MEM on amorphous, polycrystalline or rough surfaces with crystal sizes or roughness near the resolution limit of LEEM. Such surfaces are not suitable for LEEM because the intensity is backscattered over a wide angular range instead of being concentrated in a few diffracted beams. Roughness is more serious in MEM and VPEEM than in SEEM because the electrons have lower emission energies and are, therefore, more sensitive to the field distortions caused by the roughness.

The question whether or not rough surfaces are better studied by conventional scanning electron microscopy (SEM), which does not require the high field needed in cathode lens microscopy, cannot be answered in a straight forward manner. The field distortions caused by the surface roughness not only have the detrimental effect of image distortions which make image analysis difficult but also the beneficial effect of enhancing fine surface features which do not produce enough contrast in SEM. An example is the microstructure of two- and three-dimensional Cu silicide on Si(111).

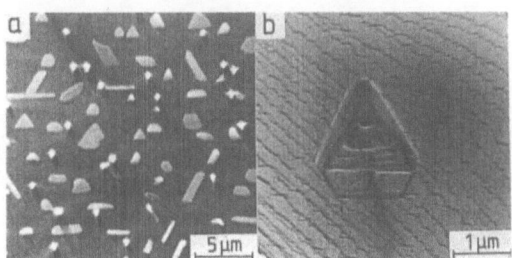

Fig. 2   Comparison between emission and reflection microscopy. Copper silicide on Si(111). a) VPEEM image, b) LEEM image, E = 4 eV.

The SEM image of such a surface [12] is very similar to the low magnification PEEM image shown in Fig. 2a [13]. No microstructure is seen in the crystals and on the flat surface between them. In contrast, the higher magnification LEEM image of Fig. 2b shows that the apparently structureless surfaces are highly stepped. The step contrast is, however, not so much caused by field distortions but mainly by interference effects between the waves reflected from the terraces bordering the steps. The high intensity, the monochromaticity and the additional information which can be extracted from the well-defined wave length make LEEM the preferred imaging mode whenever it can be used.

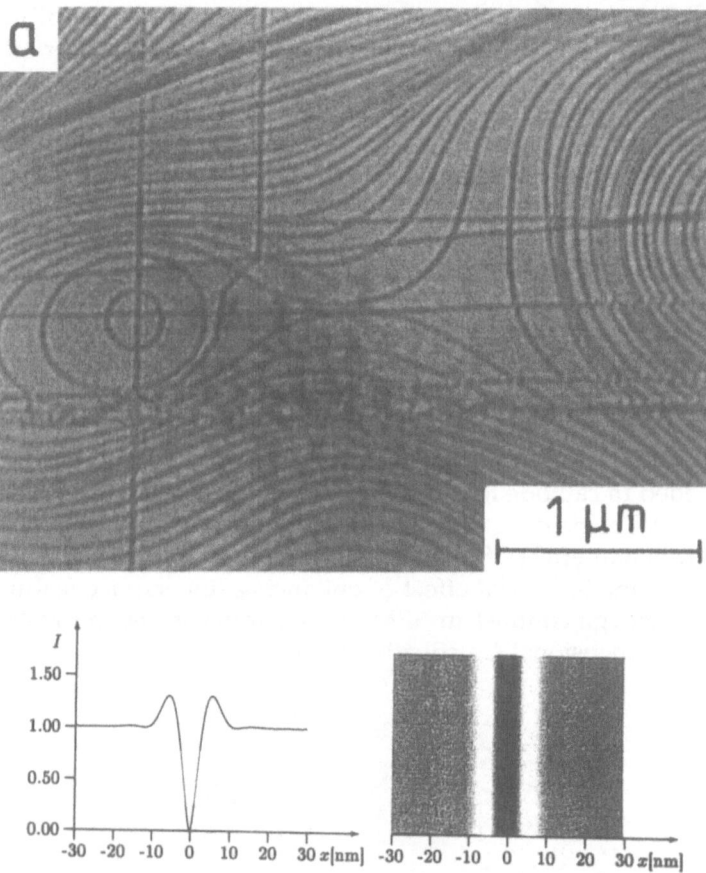

Fig. 3 a) LEEM image of Mo(110) surface with monoatomic steps, 14 eV; b) Wave- optically calculated intensity distribution and c)intensity profile at monoatomic step.

The lateral and normal resolution limits, the information depth and the contrast modes of LEEM have been discussed so often (see, for example refs. [8, 9, 14]) that only a few remarks will be made concerning the wave-optical aspects of the imaging process. A monoatomic step is a good object for testing resolution and contrast. Fig. 3a shows a LEEM image of monoatomic steps on a clean Mo(110) surface [1], taken at 14 eV which is an energy at which the phases of the waves reflected from the terraces bordering the step differ by π. The electron source was a <310> oriented field emitter operated at 25 kV and at room temperature so that the energy width was about 0.25 eV, the objective was an electrostatic triode. The resolution calculated by ray optics for this situation is about 10nm [15]. Only the first interference fringe of the interference figure can be seen. The question arises of what can be achieved by improving the resolution of the objective lens, for example with the magnetic triode used in the instrument described in Sect. 2 which has a resolution limit of about 4 nm at $E_p = 10$ eV, $\Delta E = 0.3$ eV, $U_o = 20$ kV [16]. Fig. 3b shows the intensity distribution calculated wave optically in a manner similar to that used in transmission microscopy with the following parameters: $U_o = 18$ kV, $E_p = 10$ eV, $\Delta E = 0.5$ eV, $\Delta U_o/U_o = \Delta I/I = 1 \times 10^{-5}$, illumination aperture in high voltage space 0.15 mrad. Only the first order interference fringe is visible; the profile plot (Fig. 3c) shows a very weak indication of the second fringe, too weak to be observable. Reduction of the energy width from 0.5 eV to 0.3 eV to 0.1 eV improves the resolution from 5.8 nm to 4.5 nm to about 3 nm at 10 eV. This illustrates clearly the well-known dominating effect of the chromatic aberration in LEEM and the necessity of its reduction by correcting elements such as an electron mirror or by reducing the energy width of the incident beam. However, even reduction of ΔE to 0.1 eV is not sufficient to reveal more interference fringes at a surface step unless current and voltage stability are improved significantly [17].

Except for some improvement in resolution and a considerable increase in ease and reliability of operation, there has been no major progress in LEEM in recent years. In some respect, there has actually been some deterioration in the quality of the final images because photographic recording has generally been replaced by electronic recording. In spite of the improvements in tube and CCD video technology, the signal/noise ratio is still significantly better in the photographic recording used in the past. The efforts in cathode lens microscopy were rather directed towards spectroscopic imaging which will be discussed next.

4. Energy-filtered imaging

Energy filtering always improves resolution as seen above. This is true not only for LEEM but also for SEEM, whether excited by primary electrons or by photons as in the case of threshold CPEEM, for example [3]. For AEEM and CPEEM it is a necessity in order to eliminate all but the Auger electrons or core photo electrons characteristic for the species whose distribution is to be imaged. Filtering allows an increase in the angular aperture α used in imaging because the chromatic aberration is $\delta_c \approx 2\sin\alpha \cdot \Delta E/F$ (F field strength). However with increasing electron energy E the spherical aberration $\delta_S \approx \sin^3\alpha \cdot E/F$ becomes increasingly important which makes it necessary to use smaller and smaller apertures with increasing E if a certain resolution is to be maintained. This causes intensity and signal/noise problems so that the quality of a system is no longer determined by resolution alone but also by its transmission $T = \sin^2\alpha \cdot \Delta E_F/\Delta E$ where ΔE_F is the energy window of the filter and ΔE the energy width of the spectral feature used for imaging. Depending upon the relative importance of transmission T and resolution δ several "quality factors" $Q_n = T^n/\delta^2$ can be defined. Q_n can be maximized with respect to the aperture α for n = 0,1 and 2 [18]. The case n = 0 is appropriate for LEEM where intensity is no problem, the case n = 1 may be appropriate for SEEM and n = 2 applies when as much intensity has to be collected as possible which is the case in AEEM and CPEEM. If the aberrations of the accelerating field dominate the aberrations of the cathode lens as is the case in any good lens then the optimum aperture is given by $\sin^2\alpha_2 \approx \Delta E/E$, the transmission $T_2 \approx (\Delta E^2/eU_oE)\,(\Delta E_F/\overline{\Delta E})$ and the resolution $\delta_2 \approx (\Delta E/F)\,(2\Delta E/E)^{1/2}$. If the filter slits are set such that $\Delta E_F = \Delta E \equiv \Delta E = 1$ eV which applies to many photo electron lines then for F = 6.7 kV/mm $\delta = 50$ nm and T = 5 % or $\delta = 20$ nm and T = 1 % can be reached for E = 20 eV or 100 eV, respectively, (in the homogeneous field). Real cathode lenses have lower quality factors than the homogeneous field.

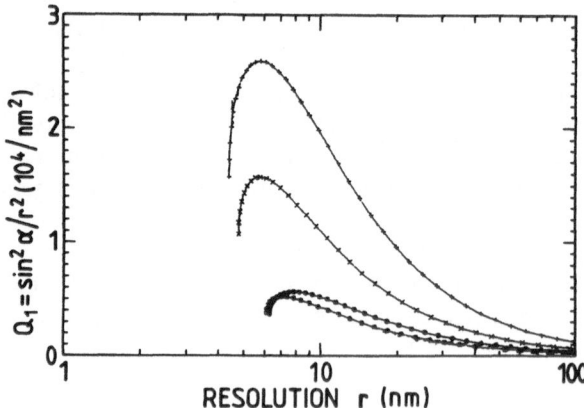

Fig. 4 Quality factor $Q_1 = T/\delta^2$ vs δ of homogeneous field and various cathode lenses for $U_o = 15$ kV, E = 20 eV, $\Delta E = 0.5$ eV (T transmission, δ resolution).

Calculations of Q_1 [19] for the cathode lenses described in ref. [16] show (see Fig. 4) that the values Q of the magnetic triode and the electrostatic tetrode are typically by a factor of 2 and 4, respectively, smaller than that of the homogeneous field. Therefore, for low intensities larger, non-optimal α´s may have to be used in order to increase transmission for acceptable signal/noise (S/N) or signal/background (S/B) ratio at the expense of resolution and low electron energies should be chosen wherever possible.

Estimates for photon flux densities of 10^{22} photons/m²s for the example E = 100 eV, T = 1 % and δ = 20 nm mentioned above indicate, nonetheless, that in favorable cases CPEEM imaging with acceptable S/N or S/B ratios and image acquisition times should be possible [18]. The example discussed was Ag on Si(111) which forms a monolayer and flat (111) oriented three-dimensional crystals. The Ag crystals should be recognizable within an image acquisition time of 1 s according to the Rose criterion $\Delta S \approx 5N$, where ΔS is the signal difference between crystal and surrounding, while for the Ag monolayer 10 s should be sufficient.

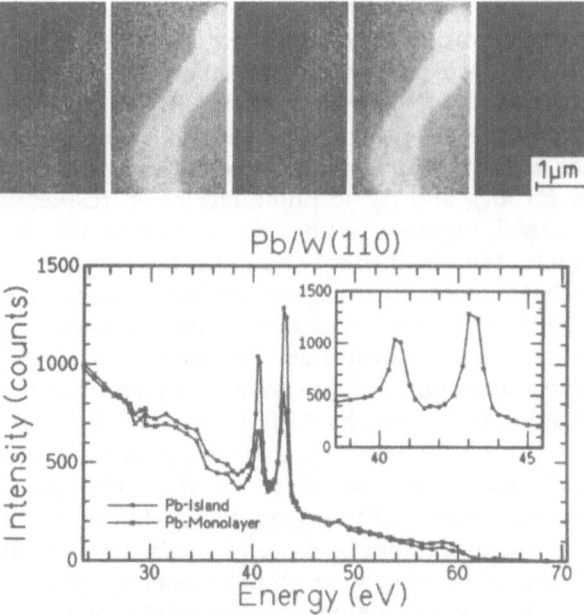

Fig. 5   Spectroscopic CPEEM images and spectra of Pb crystals and Pb monolayer on W(110). hv = 60.7eV. The energies at which the images were taken are from left to right 39 eV, 40.5 eV, 41.5 eV, 43 eV and 44.5 eV. For more information see text.

Quantitative analysis in the submonolayer range requires much longer integration times. For example for 10 % coverage accuracy near one monolayer 100 s are needed. Unfortunately, these estimates could not be checked up to now because the available undulator does not have enough brilliance at the energy needed for efficient ionization of the Ag 3d shell (> 300 eV). Therefore, Pb on W(110) was used. Pb also forms a monolayer and flat three-dimensional crystals on this surface. It has $5d_{3/2}$ and $5d_{5/2}$ ionization energies of 20.2 eV and 17.6 eV. The ionization cross-sections of these levels are high at 50 - 55 eV, the energy range in which the available undulator, the BESSY TGM5 beamline, has its maximum brightness with the 800 lines/mm grating and a gap width of 50 - 55 mm. At low magnification (1000x) image acquisition was possible with video rate, at high magnification (10000x) image acquisition times of about 30 s were necessary for sufficient S/N ratio.

Fig. 5a shows a few spectroscopic images from a series of images taken in 0.25 eV and 0.5 eV steps with an acquisiton time/image of 20 sec and Fig. 5b the XPS spectra from 0.4 μm^2 and 0.8 μm^2 areas on the crystal and the monolayer, respectively [20]. The best 15%/85% resolution at the crystal edges observed was 40 nm, a value which very likely can be improved by improved working conditions (vibrations, AC fields etc). It also has to be kept in mind that the crystal edges are not vertical but slanted. The energy resolution in these spectra as judged by the FWHM of the Pb 5d peaks after background subtraction is about 0.5 eV. The usefulness of this imaging mode for chemical identification has already been demonstrated by imaging a mixture of Pb and Ag islands with Pb $5d_{5/2}$ and Ag 4d photo electrons, respectively [20]. These results were obtained together with others in two two-week beam time periods for which the instrument had to be transported and set up. Surprisingly, in both cases no transport problems were encountered in the electron optics but only in the accessories such evaporators and electronics. Efficient operation of the instrument requires a much larger ratio of measurement time to transport and set-up time. This problem can be avoided if spectroscopic imaging is realized by AEEM, although this seriously limits the possibility of identifying the bonding state via chemical shifts.

Electron-excited AEEM can be done in the laboratory without the limitations of the available photon energy range, but with the disadvantage that the emission energy can not be chosen by varying the photon energy. In AEEM, the slow secondary peak is much more intense than in CPEEM so that operation at the low energies which is desirable from the point of view of transmission is undesirable from the point of view of the S/B ratio. In general Auger electron energies above 40 - 50 eV should be used. For Ag the $M_{45}VV$ doublet at 351 eV and 356 eV is very strong and the background relatively low for AES (S/B $\approx$ 1:2.5). The high emission energy (from the point of view of transmission) makes Ag a good test system for the possibilities of AEEM. The best Ag island signal to environment background ratio can be expected for excitation energies just below the ionization energy of the Si 1s shell (1839 eV). In the experiment to be decribed briefly below a primary energy E_p

of 2450 eV was used inadvertently, so that proper choice of E_p should lead to better results.

In order to reach in AEEM with 350 eV electrons the same resolution as with the 100 eV electrons discussed in the CPEEM case above ($\approx 20\,nm$ in the homogeneous field approximation [18]) an energy window $\Delta E_F \approx 1.5$ eV should be chosen which gives with $\sin^2\alpha_2 = \Delta E/E$ and $\Delta E_F/\overline{\Delta E} = 1.5/3.5$ a transmission $T = 0.2$ %. The lower transmission compared to the 100 eV situation in CPEEM is overcompensated by the higher electron-current density. After taking into account the other factors which determine the Auger electron yield (ionization cross section, Auger transition rate, backscattering factor, background) the final result of the comparison of CPEEM and AEEM shows that there is no difference in the sensitivity of the two imaging modes, a t least in this particular example. Fig. 6a shows some of the first AEEM images of a Ag crystal on Si(111) and Fig. 6b an Auger electron spectrum from a 1.5 μm^2 area on the Ag crystal obtained from a series of such images [21]. The acquisition time/image was 20 sec.

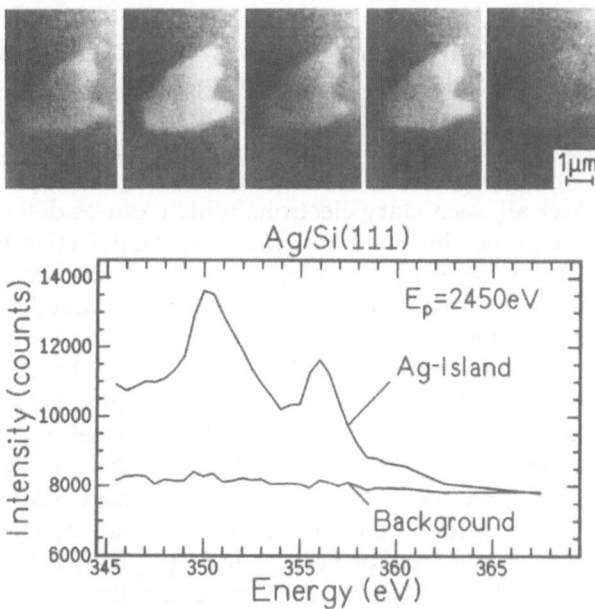

Fig. 6   AEEM images and spectrum of on Ag crystal on Si(111). E ≈ 2.5 keV. The images were taken with 346 eV, 350 eV, 354 eV, 356 eV and 360 eV electrons. For more information see text.

The examples of Figs. 5 and 6, which are the first attempts of spectroscopic imaging with a cathode lens and a beam divider show already clearly the potential of CPEEM and AEEM for chemical imaging. With a brighter undulator source as available at ELETTRA or ALS improvements in image acquistion time of at least an order of magnitude may be expected, as well as better energy and spatial resolution. CPEEM is, in principle, superior to AEEM because of the possibility of imaging different bonding configurations of a given species via chemical shifts, because of the much lower background, because of the lower load on the specimen and possibly because of the lower probability of charging of insulating specimens. AEEM, nevertheless, has its place in chemical imaging because it does not require a synchrotron but can be done in the laboratory whenever no characteristic LEED can be obtained.

5. Magnetic imaging

Finally, a comparison will be made between two methods which allow imaging of the magnetization (M) distribution in the surface region with the instrument described in Sect. 2, X-ray magnetic circular dichroism photo electron emission microscopy (XMCDPEEM) [5] and spin-polarized low energy electron microscopy (SPLEEM) [6]. These two methods are only apparently competing. XMCDPEEM is well suited for polycrystalline or amorphous materials and is chemically specific but has poor lateral resolution, while SPLEEM is ideally suited for single crystalline samples and has high resolution but does not give chemical information. The low resolution of XMCDPEEM is a consequence of the low ionization cross-sections of the core levels which show a large MCD effect (e.g. Fe, Co, Ni 2 p). This makes it necessary to collect all secondary electrons which causes deterioration of the resolution because of the chromatic aberration of the objective lens. If the core photo electrons or the Auger electrons resulting from the core level ionization are used for imaging [22], then the resolution can be improved, in principle, but only by accepting extensive image acquisition times. Until high brilliance circular polarized beam lines are available, total MCD yield microscopy is the preferred technique. Fig. 7 shows as an illustration of this imaging mode the M distribution in the near surface region of a steel specimen [23]. The image is the difference image $I\sigma_{3/2}^{-} - I\sigma_{1/2}^{-}$ taken with left circular polarized light at the BESSY SX-700-3 beam line in an image acquisition time of 30 sec/image, $I\sigma_{3/2}^{-}$ and $I\sigma_{1/2}^{-}$ referring to total yield images excited with 707.5 eV and 720.5 eV photons for $Fe2p_{3/2}$ and $2p_{1/2}$ ionization, respectively. The best resolution achieved during the first beam time was around 500 nm. There is certainly still room for significant improvement but it appears unlikely that XMCDPEEM will ever approach the high resolution and short image acquisition times which can be achieved with SPLEEM.

SPLEEM differs from ordinary LEEM mainly in the electron gun which is a Cs-oxide activated GaAs single crystal from which spin-polarized electrons are excited by circular-polarized light. Magnetic objective lenses usually have stray fields of the order 1 to 10 Gauss at the specimen position. In order to be able to work at zero magnetic field, an electrostatic objective, preferrably a tetrode lens, is used. Combining SPLEEM with spectroscopic imaging (AEEM and CPEEM) is, therefore, not straightforward and has not been attempted yet in one and the same instrument.

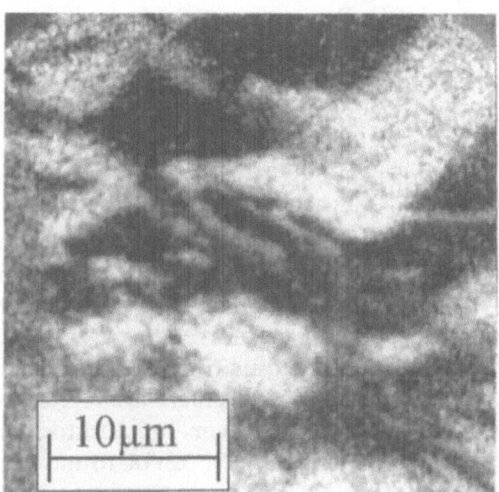

Fig. 7 XMCDPEEM image of steel. For explanation see text

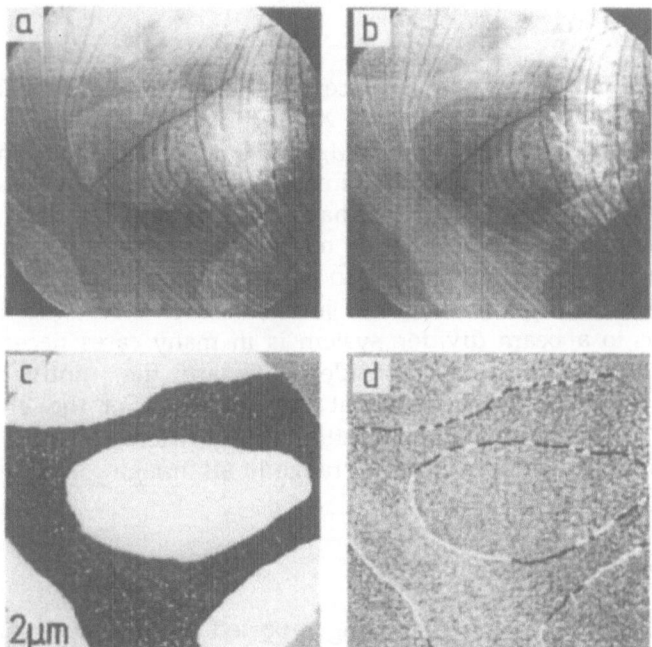

Fig. 8 SPLEEM imaging of a 8 monolayer thick Co film on W(110). Electron energy 1.5 eV. For explanation see text.

An important component of a SPLEEM instrument is also the polarization (**P**) manipulator which allows the rotation of the polarization P parallel, antiparallel or normal to the magnetization **M** in order to maximize contrast which is proportional to **P·M** [24]. Pure magnetic contrast can be obtained by subtracting two images taken with opposite **P** directions. In systems with uniaxial magnetic anisotropy either the domains or the domain walls can be imaged by rotating **P** parallel or perpendicular to **M**, respectively. This is illustrated in Fig. 8 which shows in a) and b) the individual images taken with **P** parallel and antiparallel to **M**, in c) the difference between these two images and in d) the difference image for P perpendicular to **M** [25] . If no subtraction is made, the image shows both magnetic and microstructure contrast and, thus, allows the correlation of magnetism with microstructure.

This is one of the mayor advantages of SPLEEM over other magnetic imaging techniques. Other advantages are the easy correlation with crystal structure via the LEED pattern, the short image acquisition time of the order 1 sec and less, the high resolution or the ease with which any **M** direction may be determined. Most of these advantages, however, are limited to flat, highly orientated crystalline samples. More on SPLEEM can be found in recent reviews of this still young subject [8,26].

6. Summary

In this review of recent advances in surface imaging with a cathode lens the emphasis was on LEEM, AEEM, CPEEM, XMCDPEEM and SPLEEM. Other important imaging modes such as VPEEM, MEM or SEEM have only been briefly touched. There are several other imaging or excitation modes which can be conceived and have been tried but have not reached much signifcance. The emission imaging modes, of course, do not need a beam divider, but their combination with reflection modes gives so much complementary information that the instrumental complication is well justified. Vice versa, the addition of an energy filter to a beam divider system is in many cases necessary for chemical characterization. These considerations are the motivation for further instrument developments aimed at the correction of the chromatic aberration of the objective lens and at better energy filters. If these efforts succeed, a resolution in the sub-nanometer range in all imaging modes may be expected within the next five years [27].

7. Acknowledgements

The work on cathode lens surface imaging reported in this work has been supported over the years by US Navy Independent Research Funds, by the Deutsche Forschungs-gemeinschaft, by the Volkswagen-Stiftung, by the Bundesministerium für Forschung und Technologie and by IBM Independent

Research Funds. Many other collaborators than those listed in the references have contributed to the success of this project.

8. References

1. Telieps, W. and Bauer, E. (1985) An analytical reflection and emission UHV surface
   electron microscope, *Ultramicroscopy* **17**, 57-66
2. Veneklasen, L.H. (1991) Design of a spectroscopic low energy electron microscope,
   Ultramicroscopy **36**, 76-90
3. Tonner, B.P. and Harp G.R. (1988) Photoelectron microscopy with synchrotron
   radiation, *Rev. Sci. Instrum.* **59**, 853-858
4. Kiene, M., Franz, T., Wurm, K. and Bauer, E. (1993) Direct surface imaging with
   synchrotron radiation-generated characteristic photoelectrons, *BESSY Ann. Rep.*,
   478-480
5. Stöhr, J., Wu, Y., Hermsmeier, B.D., Samant, M.G., Harp, G.R., Koranda, S.,
   Dunham, D. and Tonner, B.P. (1993) Element-specific magnetic microscopy with
   circularly polarized X-rays, *Science* **259**, 658-661
6. Altman, M.S., Pinkvos, H., Hurst, J., Poppa, H., Marx, G. and Bauer, E. (1991)
   Spin polarized low energy electron microscopy of surface magnetic structure,
   MRS Symp. Proc. **232**, 125-132
7. Veneklasen, L.H. (1992) The continuing development of low-energy electron microscopy for characterizing surfaces, *Rev. Sci. Instrum.* **63**, 5513-5532
8. Bauer, E. (1994) Low energy electron microscopy, *Rep. Prog. Phys.*, 895-938
9. Bauer, E. (1995) Low energy electron microscopy, in S. Amelinckx, D. Van Dyck,
   J. Van Landuyt and G. Van Tendeloo (eds.) *Handbook of Microscopy*,
   VCH Verlagsgesellschaft, Weinheim, to be published
10. Tromp, R. M. and Reuter, M.C. (1991) Design of a new photoemission/low-energy
    electron microscope for surface studies, *Ultramicroscopy* **36**, 99-106
11. Grzelakowski, K., Duden, T., Bauer, E., Poppa, H. and S. Chiang (1994) A new
    surface microscope for magnetic imaging, *IEEE Trans. Magnetics*, **30**, 4500-4502
12. Daugy, E., Mathiez, P., Salvan, F. and Layet, J.M. (1985) 7x7 Si(111)-Cu

interfaces: combined LEED, AES and EELS measurements, *Surf. Sci.* **154**, 267-283

13. Mundschau, M., Bauer, E., Telieps, W. and Swiech, W. (1989) Initial epitaxial
growth of copper silicide on Si(111) studied by low-energy electron microscopy and
photoemission electron microscopy, J. Appl. Phys. **65**, 4747-4752

14. Bauer, E. and Telieps, W. (1988) Emission and low energy reflection electron
microscopy in A. Howie and U. Valdre (eds.), *Surface and Interface Characteri-*
zation by Electron Optical Methods, Plenum Publishing Corporation, New York,
pp. 195-233

15. Bauer, E. (1985) The resolution of the low energy electron reflection microscope,
Ultramicroscopy **17**, 51-56 and references therein

16. Chmelik, J., Veneklasen, L. and Marx, G. (1989) Comparing cathode lens
configurations for low energy electron microscopy, *Optik* **83**, 155-160

17. Müller, T. (1995) Image formation in LEEM, M.S. Thesis, TU Clausthal

18. Bauer, E. (1991) The possibilities for analytical methods in photoemission and
low-energy microscopy, *Ultramicroscopy* **36**, 52-62

19. Chmelik, J. (1991), unpublished, see Bauer, E. (1991) Ultrahigh vacuum surface
electron microscopy with a cathode lens, *Inst. Phys. Conf. Ser. No 119: Sect. 1*,
IOP, Bristol, pp. 1-12

20. Koziol, C., Lilienkamp, G., Schmidt, T., Franz, T., Kachel, T. and Gudat,
W. (1994), Spectroscopic imaging with synchrotron-radiation generated
characteristic photoelectrons using a low energy electron microscope (LEEM),
BESSY, Ann. Rep., pp. 469-471

21. Koziol, C., Schmidt, T., Lilienkamp, G. and Bauer, E. (1995), to be published

22. Schneider, C.M., Holldack, K., Kinzler, M., Grunze, M., Oepen, H.P., Schäfers, F.,
Petersen, H., Meinel, K. and Kirschner, J. (1993) Magnetic spectromicroscopy from
Fe(100), *Appl. Phys. Lett.* **63**, 2432-2434

23. Koziol, C., Lilienkamp, G., Schmidt, T., Bauer, E., Kachel, T. and Gudat, W.
(1995), to be published

24. Duden, T. and Bauer, E. (1995) A compact electron-spin-polarization manipulator,

Rev. Sci. Instrum. **66,** 2861-2864

25. Bauer, E., Duden, T., Pinkvos, H., Poppa, H. and Wurm, K. (1995), LEEM studies

of the microstructure and magnetic domain structure of ultrathin films,
J. Magnetism Magn. Mat., to be published

26. Bauer, E. (1995) Spin-polarized low energy electron microscopy, in S. Amelinckx,

D. Van Dyck, J. Van Landuyt and G. Van Tendeloo (eds.) *Handbook of*
Microscopy, VCH Verlagsgesellschaft, Weinheim, to be published

27. Rose, H. and Preiszkas, D. (1992) Outline of a versatile corrected LEEM, *Optik* **92,**

31-44; (1994) *ICEM* **13,** pp. 197-198

SOME SPECTROMICROSCOPY DEVELOPMENTS AT BESSY

R. FINK, M. R. WEISS, V. WÜSTENHAGEN, P. VÄTERLEIN,
AND E. UMBACH

*Universität Würzburg, Experimentelle Physik II, Am Hubland,
D-97074 Würzburg,Germany*

Introduction

Spectromicroscopy and microspectroscopy have attracted considerable attention during the past five years, and technical concepts to achieve high spatial resolution using the information from different electron spectroscopies have successfully been installed at various synchrotron sources [1-5]. Two different concepts of photoelectron microscopes have been realized using scanning and imaging techniques, respectively, both of which have proven their advantages in the study of fundamental as well as applied research problems. Most concepts were presented during this workshop and shall therefore not be described here.

In this paper we review two spectromicroscopy projects using BESSY (Berlin/Germany) as photon source. During the past few years the Photon Induced Scanning Auger Microscope (PISAM) has been installed, which operates successfully with moderate lateral resolution down to about 3 mm. The installation of a focusing ellipsoidal mirror is presently under way. This improved PISAM-II instrument will allow lateral resolutions down to at best 200 nm for the C1s edge. The basics of this spectromicroscope will be described shortly, and some recent experimental results will be presented. In April 1995 a new research project has started to develop an imaging microscope based on the LEEM idea by E. Bauer and coworkers at TU Clausthal (Germany) [5]. In addition to the LEEM mode the new instrument will be operated as a XPEEM with ultimate resolutions of less than about 5 nm, which is achieved by correction of chromatic and spherical aberrations. The layout of this new instrument is presented and discussed shortly.

R. Rosei (ed.),
Chemical, Structural and Electronic Analysis of Heterogeneous Surfaces on Nanometer Scale, 93–102.
© 1997 *Kluwer Academic Publishers.*

The Photon Induced Scanning Auger Microscope - PISAM

DESCRIPTION OF THE APPARATUS

The PISAM [6] was developed during the past years and has shown ist potential for spatially resolved chemical analysis of surfaces by means of Auger spectroscopy with a typical spatial resolution of 3 - 4 μm [6,7] and energy resolution of about 0.5 eV. It can also be operated in the photoemission mode but with reduced spatial (20 μm, due to diffraction) and energy resolution (approx. 2 eV, due to the width of the quasimonochromatic undulator radiation). This microspectroscope in ist present configuration essentially consists of a set of apertures which reduces the beam of the first BESSY undulator W/U-1 laterally, thus creating a fine spot of X-rays (see fig. 1).

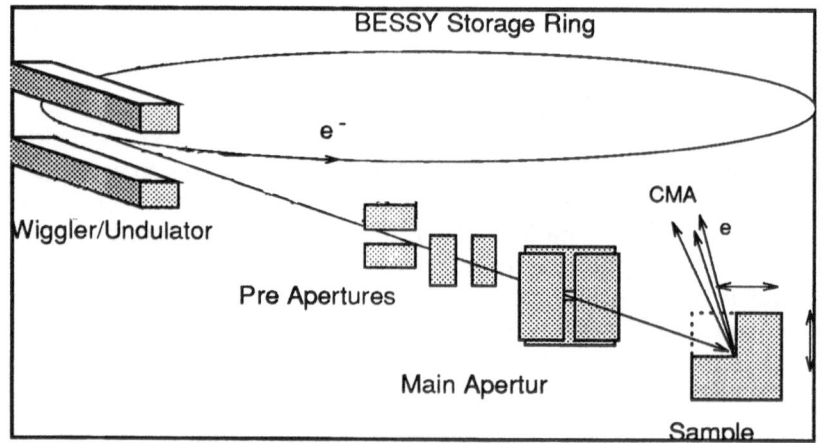

Fig. 1: Schematic drawing of the initial PISAM apparatus.

The apertures can be closed down to a few micrometers limited by the spreading of the radiation behind the aperture due to diffraction. This surprisingly simple setup operates successfully because of the little divergence of the undulator beam (approx 0.15 mrad) and because of its high flux (typical brilliance of 5×10^{15} Phot/sec(0.1mrad)^2 1% BW).

The excited photo- and Auger-electrons from the sample are recorded by a spherical sector analyzer with a typical energy resolution of about 0.5 eV. By identifying the features of the electron energy distribution one can draw conclusions about the chemical composition of the sample, thus employing the PISAM a microprobe for chemical information from the sample surface. Three modes of measurements are possible: photoelectron and Auger electron spectroscopy from a small spot on the sample („microspectroscopy"), or intensity scans of selected energy windows along a line on the sample, or across a 2D-region of the surface („spectromicroscopy").

At present, the spatial resolution of the PISAM is limited only by diffraction at the aperture yielding 20 µm at hν = 50 eV and 3-4 µm at hν = 300 eV [7]. The flux through the smallest useful aperture is still 10^{11} phot/sec (for hν = 300 eV) and 10^{12} phot/sec (for hν = 50 V), respectively. A further reduction of the aperture size would in principle reduce the spot size if the distance between aperture and sample is reduced accordingly, but is not possible because of the spatial restrictions for the emitted electrons.

Experiments

CHEMICAL MAPPING OF CIGS BASED SOLAR CELLS

Chemical mapping of $Cu(In,Ga)Se_2$ (CIGS)microstructures is chosen as example in order to demonstrate a main purpose of the PISAM. Ultrathin $Cu(In,Ga)Se_2$ films are very promising absorber materials for low cost and high efficiency solar cells. Their efficiency is found to correlate with the sodium content segregating from the float glass substrate through the Mo back electrode into the CIGS absorber film. A laser cut technique is regarded as a possibilitypresently used for microstructuring and connecting the single absorber cells. Chemical analysis of the microstructure and its correlation with the performance of such a microstructure in operation is therefore of particular interest.

particular interest.

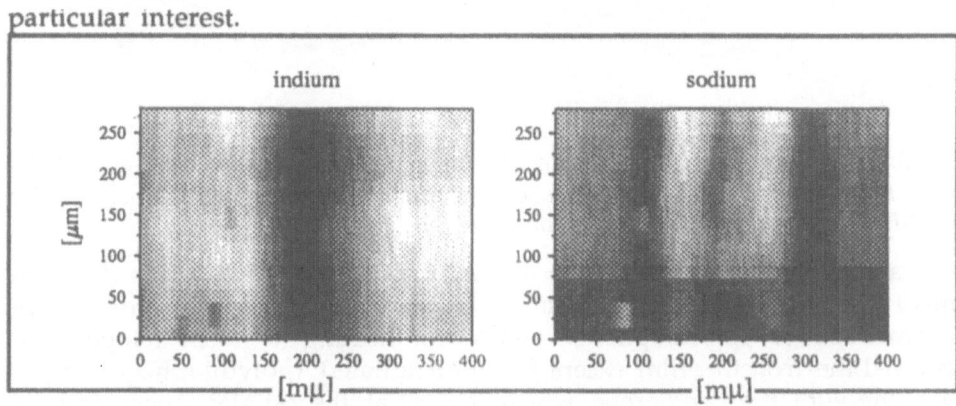

Fig. 2: Intensity maps for indium (left) and sodium (right) in the vicinity of a laser cut. The detected area is about 400 x 280 mm².

Fig. 2 shows 2D greyscale intensity images for sodium and indium derived from corresponding valence band structures in the vicinity of a (vertical) laser cut through the Mo film (the cut was made prior to the growth of the $Cu(In,Ga)Se_2$ absorber). It clearly demonstrates increased Na content near the laser cut as well as varying In content in the same area. The stoichiometry of the film varies significantly with respect to areas far from the laser cut. Consequently, new structuring techniques or substrate modifications might be necessary in order to improve the spatial distribution of the sodium content in the absorber material.

WRITING AND ANALYSIS OF MICROSTRUCTURES

Several experiments have been performed on the analysis of metal microstructures on semiconductors and of microstructured organic films on metals. Moreover the production of microstructures by „direct writing", i.e. by photolytic synthesis or decomposition, has been tested.

Polymer Microstructures

Due to the high photon flux, structuring by bond breaking or by stimulated bond formation, e. g. polymerization of organic compounds such as thiophene, is feasible. This latter class of compounds has recently gained interest due to its model character for applications in molecular electronics. In order to produce polythiophene microstructures we condensed a thin film (several nm thickness) of thiophene monomers on Au/SiO_2 at LN_2 temperature.

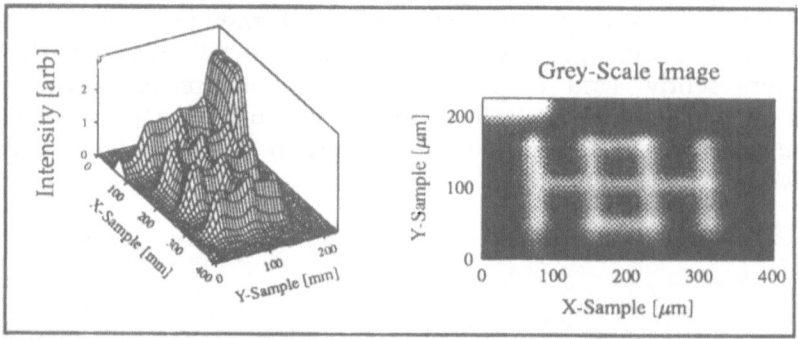

Fig. 3: Thiophene structures recorded as Auger intensity distribution at 152 eV. High intensity indicates the presence of sulfur as polythiophene produced by photolytic polymerization of monothiophene.

By scanning the X-ray beam (hv = 45 eV, exposure time 4 s) across the film preset structures could be written since the exposed areas were photolytically polymerized to polythiophene. Subsequent heating to room temperature removed the monomer film except for the irradiated (polythiophene) areas. Mapping with the microprobe, which was set at the S-LMM Auger peak of molecular sulfur and hence sensitive to occurrences of thiophene, clearly showed the polymerized structures [8, 9] (chemical map see fig. 3).

Direct Writing of Metallic Microstructures

Similar experiments were performed with palladium-acetate [9] and Mo-hexacarbonyl [10]. In these two cases metal microstructures were directly „written" by decomposition and desorption of the acetate or CO fragments. As an example, writing of Mo microstructures on Si will be described in brief.

Exposure of $Mo(CO)_6$ to soft X-rays adsorbed on a Si wafer leads to photolytic dissociation with desorption of most of the CO ligands. Therefore $Mo(CO)_6$

multilayers were irradiated for 7 min using an aperture of about 30×30 μm². The measurement of photoelectron spectra took only 10 seconds. Fig. 4 (left part) compares such valence band spectra for the X-ray "exposed" and "not exposed" sample for element identification. The peak at about -2 eV binding energy of the not-exposed sample (dashed line, see arrow) is assigned to Mo 4d-derived valence orbitals and the peaks at -8.5 and -11.8 eV (see arrows) to the CO-derived 5s / 1p and 4s molecular orbitals, respectively. On the right, a line scan across the irradiated spot is shown for three energy windows recording electron intensity at -10 (A), -6 (B) and +7 eV (C) binding energy. 2D writing of Mo microstructures was demonstrated on a larger scale with significantly shorter exposure times (10 s) and recording times of 20 ms per pixel [10].

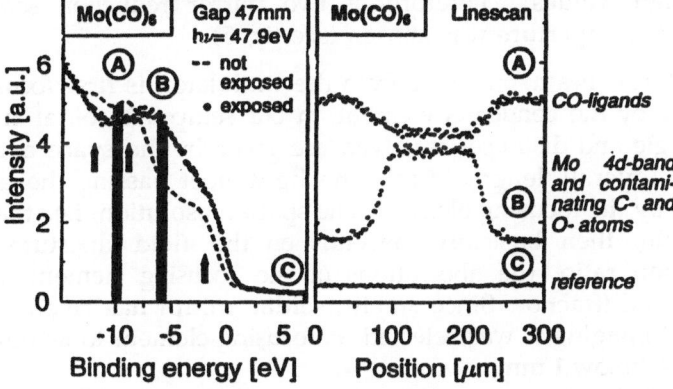

Fig. 4: Photoelectron spectra (left) for irradiated (solid dots) and non-irradiated (dashed line) $Mo(CO)_6$ adsorbed on a Si substrate. The right-hand side shows line scans using electron energy windows at binding energies indicated by grey bars and capital letters on the left.

These experiments prove the ability of the instrument to combine direct writing of structures with quasi-simultaneous in-situ analysis. Using both, writing metallic as well as organic microstructures will allow to perform in-situ preparation and analysis of polymeric microstructures and might thus open the opportunity to create small scale electronic test devices based on organic (polymeric) materials.

New PISAM Concept (PISAM-II)

The limited spatial resolution is the major disadvantage of the present PISAM instrument as compared to some of the other projects [2 - 4]. Higher spatial resolution is of course desirable for most microanalysis and structuring applications. Therefore efforts were undertaken to increase the resolution with an improved concept.

Higher resolution, without sacrificing flux, is achievable only with focusing elements. Several concepts of focusing x-rays have been developed of which the most common are: grazing incidence mirrors, multilayer mirrors, and micro zone plates. For the new PISAM instrument we favoured the concept of grazing incidence mirrors, since they

- sustain high thermal loads
- have minor band limitations (photon energies can be chosen in a wide range),
- have no chromatic aberration, and
- are relatively easy available with reasonable costs,

and our instrument uses high flux and variable photon energies. However, grazing incidence mirrors are manufacturable only with small numerical aperture at small grazing angles. In addition, the reduction ratio cannot be pushed to high values. Therefore, a two mirror reduction scheme with intermediate field aperture was chosen (see fig. 5).

The intense X-ray beam, emerging from the undulator, is first focused on the field aperture by the condenser element, in our setup a toroidal mirror. The deflection angle and distances involved are given by the space available at the beamline (overall length 16 m). The light spot passing the aperture is then reduced by the focusing element. The spatial resolution, i.e. the spot size on the sample, then basically depends on the field aperture size, the demagnification ratio, the aberrations of the focusing element, its surface quality, and on diffraction. Since an ellipsoidal mirror has no aberrations for point-to-point imaging it was selected as focusing element to achieve lateral resolution well below 1 mm.

We have singled out two concepts (demagnification $q = 15$, incident angle $\varphi = 5°$ for low cost but less resolution) and ($q = 20$, $\varphi = 6°$) which is expected to yield higher resolution but causes higher production costs. For these two parameter sets raytrace analyses were performed to find the intensity distribution in the image plane considering geometric effects. Diffraction was included by convolution of the diffraction pattern of the aperture stop at the mirror with the results of the raytracing [11].

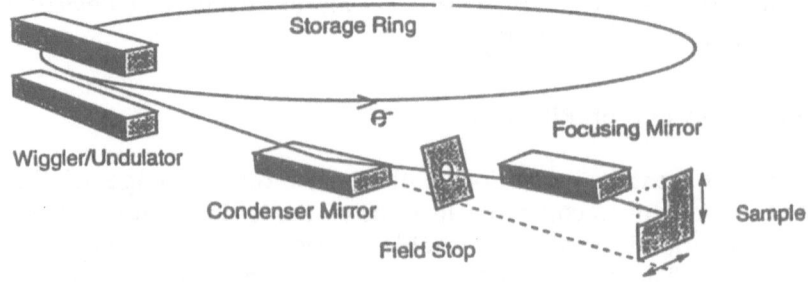

Fig. 5: Scheme of the new PISAM-II concept

The evaluation process resulted in a design with the following properties:

- focus width 0.2 - 0.3 mm
- focus height 0.9 - 1.1 mm
- photon flux of more than 2×10^{12} phot/s 1\% BW, for the smallest possible focus.

The optimization process taking into account tangential errors, micro-roughness, and diffraction as well as manufacturing and physical constraints resulted in the above described focusing concept. The PISAM-II instrument will enable high resolution micro-spectroscopy with a spatial frequency of $\leq$ 3 mm^{-1} (MTF, 11% threshold) for energies between 45 and 600 eV and an energy resolution of the detected electrons of approx 0.5 eV at very high photon fluxes ($> 2 \times 10^{12}$ phot/s 1% BW) (note that this is equivalent to photon fluxes of more than $> 2 \times 10^{18}$ phot/s 1% BW of sources with normal 1 mm^2 spot size). Consequently, very short measurement times of Auger and photoemission spectra as well as direct writing of microstructures will be possible with high spatial resolution (0.2 -1 mm).

Although the lateral resolution of the PISAM-II instrument is still one order of magnitude worse than today's best imaging spectromicroscopes, it will be of interest for "direct writing" and for the analysis of polycrystalline samples where large surface corrugations are present. In these cases the high electric fields required for high PEEM resolution will negatively influence the reachable lateral resolution. The PISAM-II is partly operating; the toroidal condenser mirror is installed and is performing as planned. The manufacturing of the ellipsoidal mirror and the hardware construction of the new setup is almost finished. The complete PISAM-II is expected operate at BESSY-I in the beginning of 1996.

The New XPEEM/LEEM Instrument

A new project of an imaging spectromicroscope has recently started based on a collaboration of the universities of Bochum, Clausthal, Darmstadt and Würzburg, the Fritz-Haber Institute in Berlin, BESSY and the LEO (former Carl Zeiss) company in Oberkochen. The concept of the new microscope is based on the idea of a low energy electron microscope (LEEM) as proposed and developed by Bauer and coworkers [5]. In comparison to present spectromicroscopes [1, 5] the new instrument focuses on minimization of chromatic aberration as well as reduction of errors arising from the large acceptance angle. These errors are corrected up to second order by additional components leading to the layout presented in Fig. 6.

In order to achieve a high photon flux as well as a large photon energy range with high spectral resolution ($> 10^4$), the microscope will be installed at a multiperiod (N> 80) U49-undulator at BESSY-II equipped with a high-flux, small-spot plane grating monochromator of the BESSY HE-PGM3 type. The photon density at the sample will be increased by an ellipsoidal grazing incidence mirror demagnifying the monochromator exit slit by a factor of 10. The illuminated spot size on the sample will be in the order of $5 \times 10 \, \mu m^2$ with a photon density of about 10^{12} photons/s·μm^2.

The emitted photo-, secondary and Auger electrons pass through the object lens and through a magnetic sector field. They are reflected by a tetrode mirror to continue on a symmetric path through the magnetic sector field and a transfer optics to the imaging electron analyzer ("Omega filter"). This is also corrected for errors up to second order and therefore allows a rather high electron energy resolution (0.1 - 0.2 eV over an energy range of 0 - 1500 eV). Magnification is achieved in the projection lens before the electrons are registered in a two-dimensional detector array.

The most important development in this instrument is the implementation of a compact stigmatically imaging 90° beam splitter with symmetric arrangement of the conical electromagnetic immersion objective lens, a mirror hexapole corrector (tetrode mirror), an electron point source, and the projection system. This setup was proposed and is further developed by Rose and Preikszas [12]. Due to its high symmetry the electron microscope does not introduce dispersion and second order aberrations at the image plane. The correction increases the resolution by a factor between four and eight and allows, according to the calculations, lateral resolution in the *sub*nanometer range. However, this would imply mechanical and technical perfections which probably cannot be realized at present.

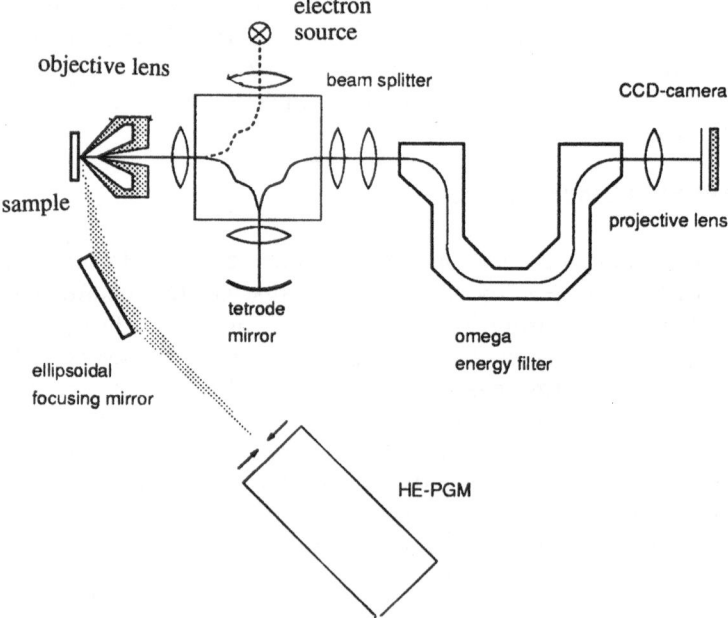

Fig. 6: Layout of the future XPEEM/LEEM at BESSY which should allow lateral resolution well below 5 nm.

In its final state this spectromicroscope should enable photoemission, Auger and X-ray absorption experiments with very high spatial or with high energy resolution, or electron microscopy and small spot LEED experiments with extremely high spatial or (modest) time resolution using the electron point source. Of course, with such an instrument many interesting microscopic questions could be tackled which cannot be answered either by conventional microscopy or by present microspectrocopes.

Finally we should mention that another XPEEM instrument with less complicated electron optics is presently developed by a Mainz/Bielefeld collaboration (Profs. Schönhense and Heinzmann) using a corrected Wien filter for energy detection and a multilayer mirror for photon focusing.

Acknowledgments

We gratefully acknowledge helpful discussions with W. Peatman (BESSY) concerning the PISAM-II concept. We are indebted to H. Petersen (BESSY), who was strongly involved in the monochromator design for the XPEEM/LEEM (he passed away on May 8, 1995). Thanks to all collaborators in the XPEEM project. Financial support was/is given by the BMFT/BMBF under contracts 05 5VSAXB, 05 5WWAXB (PISAM) and 05 644WWA (XPEEM/LEEM).

References

[1]   B. P. Tonner and G. R. Harp, „Photoelectron microscopy with synchrotron radiation", Rev. Sci. Instrum., **69** (6), p. 852 (1988). B. P. Tonner, this workshop.

[2]   C. Capasso, A. K. Ray-Chaudhuri, W. Ng, S. Liang, R. K. Cole, J. Wallace, F. Cerrina, G. Margaritondo, J. H. Underwood, J. B. Kortright, and R. C. C. Perera, „High-resolution x-ray microscopy using an undulator source, photoelectron studies with MAXIMUM", J. Vac. Sci. Technol., **A9** (3), p. 1248 (1991). G. Margaritondo and F. Cerrina, this workshop.

[3]   H. Ade, J. Kirz, S. L. Hulbert, E. D. Johnson, E. Anderson, and D. Kern, „X-ray spectromicroscopy with a zone plate generated microprobe", Appl. Phys. Lett., 56 (19), pp. 1841-1843 (1990). H. Ade, this workshop.

[4]   J. Voss, H. Dadras, C. Kunz, A. Moewes, G. Roy, H. Sievers, I. Storjohann, and H. Wongel, „A Scanning Soft X-Ray Microscope with an Ellipsoidal focusing Mirror", J. X-Ray Sci. Techn., 3, p. 85 (1992).

[5]   E. Bauer,"Low Energy Electron Microscopy", Ultramicroscopy, 36, p. 52 (1991).
      E. Bauer, this workshop.

[6]   V. Wüstenhagen, M. Schneider, J. Taborski, W. Weiss, and E. Umbach, „Concept and realization of a photon-induced scanning Auger microscope", Vacuum, **41** (7-9), p. 1577 (1990).

[7]   V. Wüstenhagen, Doctoral Thesis, Universität Stuttgart, July 1992.

[8]   M. Weiss, Diploma Thesis, Universität Stuttgart, January 1993.

[9]   P. Väterlein, M. Weiss, V. Wüstenhagen and E. Umbach,"Mask-less writing of microstructures with the PISAM", Appl. Surf. Sci., **70/71**, p. 278 (1993).

[10]   P. Väterlein, V. Wüstenhagen, and E. Umbach, „Direct writing of Mo microstructures using high brilliance synchrotron radiation", Appl. Phys. Lett. **66** (17) p. 2200 (1995)

[11]   M. Weiss, V. Wüstenhagen, and E. Umbach, „A grazing incidence mirror for a new microspectroscope using soft X-ray undulator radiation" Proceedings SPIE, p. 1 (San Diego 1994)

[12]   H. Rose, and D. Preikszas,"Outline of a versatile corrected LEEM" Optik **92** (1) p. 31 (1992); D. Preikszas and H. Rose, „Corrected LEEM for Multimode Operation" Proceedings ICEM-13 (Paris 1994) p. 197

SHEDDING LIGHT ON SURFACE REACTIONS:

PEEM, EMSI and RAM probed chemistry on solid surfaces.

H. H. ROTERMUND

Fritz-Haber-Institut der Max-Planck-Gesellschaft,
Faradayweg 4-6, D-14195 Berlin, Germany

Different submonolayer adsorbate coverages on surfaces during reactions can be visualised by several non-destructive microscopies. By using photons as the incident particles the emitted photoelectrons may be imaged and the picture will reveal differences of the work function (PEEM). If the reflected light is exploited, either an ellipso microscope for surface imaging (EMSI) or a reflection anisotropy microscope (RAM) can be utilised. Pattern formation during the catalytic CO-oxidation is observed by these methods over a wide pressure range annihilating the "pressure gap" between ultra high vacuum (UHV) and ordinary atmospheric conditions.

1. Introduction

Understanding of surface reactions is of basic interest for heterogeneous catalysis. Under certain experimental conditions, various catalysed reactions exhibit oscillations of the reaction rate. This has been known for the CO-oxidation on polycrystalline samples since the early seventies while about 10 years later a mechanism on single crystal surface was proposed by Ertl et al.

R. Rosei (ed.),
Chemical, Structural and Electronic Analysis of Heterogeneous Surfaces on Nanometer Scale, 103–129.
© 1997 *Kluwer Academic Publishers.*

Although elaborate studies had been performed information about spatio-temporal pattern formation was not available. First hints that laterally non-uniform processes take place were given by Cox et al. where scanning LEED was used with a lateral resolution of about 1 mm and time resolution of several seconds.

Due to the rather rough lateral and time resolution no pattern formation could be directly observed. To increase the lateral resolution a scanning photoemission microscope (SPM) was developed in which a simple deuterium discharge lamp was focused to a spot of about ø = 1 µm and then scanned across the surface while in each point the total yield of photoemitted electrons were counted and imaged with a computer. First results during oscillations of the CO-oxidation on a Pt(100) where published in.

Even though the lateral resolution had been improved by 3 orders of magnitude compared to the scanning LEED experiment, the time response of this new instrument on the order of seconds, was still rather slow. Since with any scanning technique "real time" imaging is virtually impossible, the design and construction of a photo-emission electron microscope (PEEM), capable of parallel imaging seemed to be the only proper solution. From the earlier investigations the design criteria were clear: a rather wide (400 µm) field of view, resolution to about 0.1 µm, but now real time imaging at least in terms of video recording. Therefore W. Engel designed the electron optics for a rather straightforward PEEM specific to our needs, which nevertheless tackles many different problems in surface science. We then constructed it as an "add on" microscope for any standard UHV-vessel. All common LEED or Auger-system can be exchanged with the PEEM.

Even the very first tests of only the objective lens revealed already a wealth of previously unseen patterns in a surface reaction. Since 1991 direct copies of this instrument are commercially available and used in a variety of research areas such as morphology studies of carbon films, diffusion of alkalis on metal substrates or of adsorbates like CO on Pt-surfaces.

For questions relevant to real catalysis the drawback of the PEEM arises from its limitation in terms of vacuum conditions. Designed to be fully ultra high vacuum (UHV) compatible, we can operate the PEEM by differential pumping through apertures of 200 μm and 1 mm in the electron optical axis at total pressures as high as 10^{-3} mbar. The remaining "pressure gap" of 6 orders of magnitude or more between UHV experiments and atmospheric catalytic conditions is inherent to nearly all tools used in surface science and any technique capable of bridging this gap will be of great value.

Optical methods have this potential in principle. Like ellipsometry, some have been used in surface science for a long time. Their sensitivity, to even submonolayer coverages have been demonstrated for many systems. Even though the zenith of ellipsometry has passed a long time ago, it is still frequently used in many industrial processes like molecular beam epitaxy for controlling the thickness of layers. During the last decade a strong desire for lateral information has created several new surface science instruments, but ellipsometry had been only used in vacuum as an integral method or, at best, in a scanning set up. Ruud Tromp (personal communication, September 1994) proposed an imaging technique using ellipsometry, while we were participating in a fire drill during the middle of a seminar I gave at IBM, Yorktown Heights. We then developed a method called "Ellipso-Microscopy for Surface Imaging" (EMSI) and subsequently a relative of it, the "Reflection Anisotropy Microscope" (RAM). After describing the different experimental approaches in some detail , I will discuss some results on the CO-oxidation on various Pt surfaces achieved by using the PEEM, RAM and EMSI.

2. Experimental set up:

All experiments were carried out in a UHV-vessel with a base pressure of 2×10^{-11} mbar. A gas inlet system with feedback - stabilisation was used to keep the partial pressures of CO and oxygen constant during the experiments. This system is described here in detail . The chamber is also equipped with an ion - gun for sputtering, facilities for LEED and Auger electron spectroscopy and two mass spectrometers, one of them differentially pumped to follow in situ the reaction rate of CO_2.

A broad spectrum of pattern formation during surface reactions were imaged with the different methods PEEM, EMSI and RAM. The first two of them could be operated simultaneously for the identical sample position. The contrast mechanism differs substantially from one method to another and will be described below.

The PEEM images differences in the work function of surfaces, for instance due to the non uniform coverage with adsorbates. Depending on the lateral resolution used, differences as small as 2 meV can be detected. It consists of three electrostatic electron lenses which are mounted on a standard 150 CF flange, interchangeable with any standard LEED or Auger system. As in all emission microscopes, the first lens works as a cathode lens. For this purpose a high voltage between the sample and the first lens electrode provides the acceleration for the photoemitted electrons up to 20 kV. By using a positive high voltage for the microscope column, the sample can be left at ground potential, allowing a standard UHV manipulator to be used with easy provisions for temperature control. One important feature of this PEEM is its rather large field of view of 400 μm diameter at 100x magnification at the designed working distance of 4 mm. A field of view with 700 μm diameter still produces excellent pictures. The magnification can be increased to about 1000x, while maintaining a resolution better than 0.3 μm. The sample is illuminated through an additional viewport with UV-photons from a deuterium discharge lamp having a cut-off energy 6.5-7.0 eV, depending on the lens material, the windows and the amount of oxygen in the path of the photons. With such excitation energies, the CO covered areas remain grey in the image, while the O covered regions on the same crystal having the highest work function appear dark. Brightness and contrast are influenced by the photon flux densities of the light source, the gain of the image intensifier, a multi channel plate (MCP), the sensitivity of the video camera among other factors. To protect the MCP, the column is pumped differentially by a turbomolecular pump. The image itself can be viewed directly from the outside and is recorded with a CCD video camera onto a S-VHS tape, which can be further analysed by computer.

With EMSI, the adsorbates and their coverages are imaged by the differences of their optical properties even at sub-monolayer thicknesses. The principle for the EMSI set up is reproduced in fig. 1.

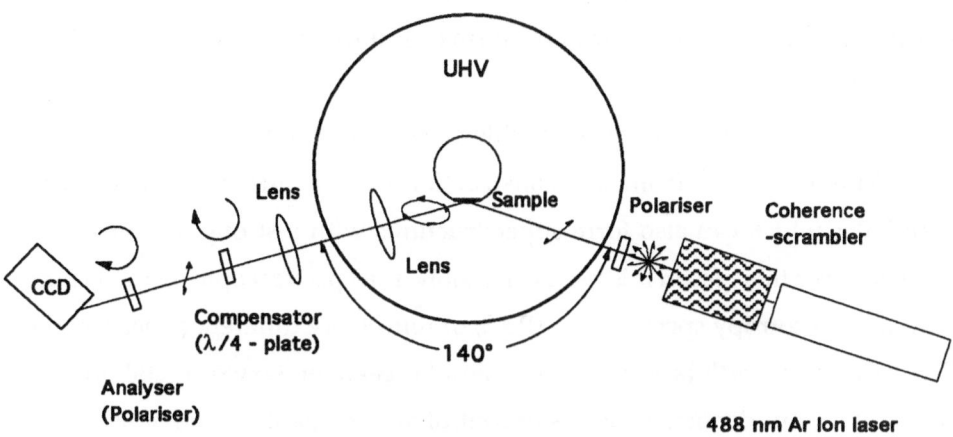

Figure 1. Ellipso - Microscope for Surface Imaging

The optical path consists of a single line (488 nm) 100 mW Ar Ion Laser. Its coherence is scrambled to reduce disturbing interference patterns in the final image, caused by dust particles or optical imperfections, by passing the light through a vibrating multimode optical fibre. The resulting small light source at the exit of the optical fibre is projected onto the sample by a collimator through a polariser (Glan-Thompson) and a viewport (silica grade glass, chosen to display small birefringence) at an angle close to $70°$ from the surface normal. The beam is reflected from the sample and thereby changes from linear polarisation into elliptical polarisation, as indicated by the ellipse in fig. 1. Two lenses were used for imaging , one inside of the UHV with a focal length of f = 60 mm, the other one located outside, with f = 50 mm. The light then passes through a quarter wave compensator plate which is turned in such a way that again linearly polarised light is created. By the following analyser its intensity is adjusted, again by turning it as indicated in fig. 1, to near zero and then measured by a CCD chip. The latter is tilted against the optical axis to reduce distortion and to increase sharpness of the image. It is further improved by the use of an image processor, which is capable of subtracting a background image in real time (Argus 20 by Hamamatsu). We normally chose an image of the homogeneously CO or O covered surface as a background, usually averaging up to 4 s of video frames. Changes of the local coverage will give rise to a signal, which in our case occur in the form of spirals, wave trains or just single reaction fronts.

RAM achieves its contrast via a possible anisotropy of reflection for polarised light. This can arise from a reconstruction of the surface itself or from adsorbates which can also form superstructures with a strong anisotropy for reflection. For this reason it is used commonly as a spectroscopic method called reflection anisotropy spectroscopy (RAS) or reflectance difference spectroscopy (RDS) during growth processes like molecular beam epitaxy or metal orga-nic chemical vapour deposition as in situ control for the quality of the films.

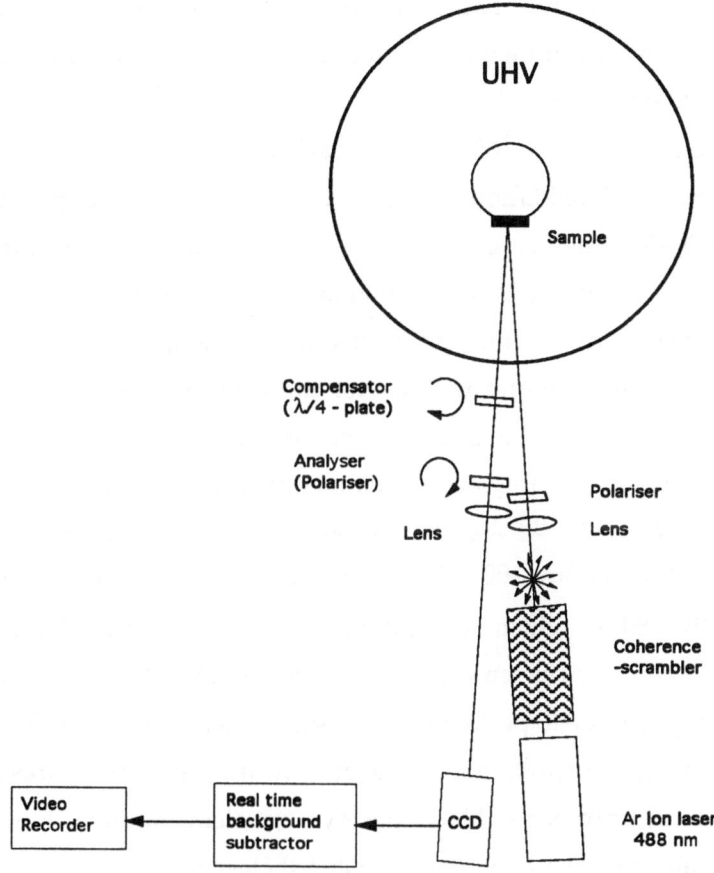

Figure 2. : Reflection Anisotropy Microscope

Our RAM as sketched in fig. 2 uses identical optical components as they are described for the EMSI, but the optical pathway is adjusted such, that there is only about 8 degrees between the ingoing and reflected beam. Ideal would be of course normal incidence to the surface. Care must be taken that the polariser is adjusted between the anisotropy axes of the surface. The specularly reflected light exits the UHV through the viewport and passes a retarder (quarter wave plate) and a second polariser acting as analyser for the polarisation state of the reflected light. Spatial resolution is achieved by a lens, allowing magnifications between 2x and 20x by which an image of the surface is projected onto a CCD chip. The same video processing unit, as with EMSI, allows image enhancement by on-line subtraction of a background. The final images are stored on S-VHS video tape.

For Pt(110) the anisotropy in reflectivity is caused by the anisotropic surface structure. During CO oxidation changes of this anisotropy, appearing as contrast in the RAM images, are related to a O or CO induced restructuring of the surface. If the adsorbate is in addition also structured in an anisotropic arrangement this would also give contrast in our RAM images.

Since completely new pressure regimes for the experiments are accessible, many additional problems have to be overcome. For instance, we already found a drastic increase of the reaction temperature, starting at 10^{-2} mbar of oxygen pressure, due to the higher production rate of CO_2. At a partial pressure for O_2 of about 100 mbar, the heat of reaction increased the temperature by 180 K, which was equivalent to more than doubling the electrical heating power. The temperature dropped on the other hand, after the reaction had stopped due to poisoning of the surface with CO by 100 K in 30 s. A different approach to temperature control is inevitable. At pressures above 0.1 mbar we have only been able to observe transient patterns due to the problems in stabilising the temperature. Nevertheless the changing coupling mechanism of the self-organisation from diffusion to thermokinetic control will now be accessible.

Of course the pumping system needs modification, since normally the pumps are designed to produce UHV conditions and are not operable for a longer period in the mbar range. For pressures from 0.05 to 100 mbar pumping was performed by a choked roughing pump only. By using high purity gases (4.7 for CO and 5.6 for oxygen) excessive contamination of the chamber was prevented and a base pressure of 10^{-9} mbar maintained without baking the system. From 10^{-6} to 1 mbar the pressure was measured capacitively by a Baratron, and at higher pressures with a pirani gauge.

The heating of the sample, usually an indirect heating filament at the back of the sample, is rather inappropriate for high partial pressures of oxygen, resulting in rapid burnouts. We switched to heating the sample from the back by irradiating it with a 24 V, 250 W halogen light bulb, shielding out all stray light. The Pt(110) sample (diameter 10 mm, thickness 1.5 mm) could easily be heated to 1200 K for preparation purposes. The sample temperature is measured by a NiCr-Ni thermocouple in a hole in the crystal.

Comparing our new methods with the PEEM the lateral resolution has to be less than with the latter, since optical microscopes are normally diffraction limited. Nevertheless they are easily sufficient for the fields of view necessary for our purposes, ranging from several hundred μm to some mm in diameter. The time resolution is basically restricted by the video system used, but in principle holds much more potential for EMSI and RAM compared to a PEEM, since increasing the signal intensity is always possible without the space charging effects which would occur in photoelectron imaging.

Since the optimum angle for the ellipsometric observation is about 70 degrees from the surface normal for most metals, the surface itself is now easily accessible for various interactions. It should be possible to follow local sputtering or evaporating patterns in real time and space.

An additional advantage of the new methods besides being independent of the working pressure, are their capability to image surfaces even for very aggressive chemical reactions, where standard UHV equipment would have only a very short life time. It is also possible to investigate interfaces between liquids and solids.

3. Results:

In this section we discuss examples from the three imaging techniques, primarily for the oscillatory oxidation of CO on Pt surfaces. In general oscillations of the reaction rate on Pt(100), Pt(110) and Pt(210) were proven to follow a common principle:

A structural phase transition of the surface induced by adsorbates, influences the sticking coefficient of oxygen, inducing a feedback mechanism. One of the systems under current investigation is the Pt(110) surface. There the proposed surface phase transition occurs between the reconstructed (2x1) - configuration realised for the clean or oxygen covered surface and the 1x1 - surface. This surface phase transition is initiated by CO adsorption which begins to occur at 0.2 monolayer coverage of CO, and is completed with 0.4 monolayer as corroborated by STM.

The reverse transition from the 1x1 to 2x1 structure takes place when the coverage of CO falls below 0.2. The most striking difference between the 1x1 - phase and the 2x1- phase is the sticking coefficient for oxygen. On the 1x1 - phase the oxygen sticking coefficient is about 0.6, but only 0.4 on the reconstructed phase. If the partial pressures and the temperature are chosen appropriately, pattern formation during the reaction might occur. These non-linear effects can be divided into several sections in parameter space, each with completely different pattern formation.

The Pt(100) surface generally produces irregular oscillation of the reaction rate and this manifests itself in the form of the patterns. The region where we have found patterns with the PEEM is called a "double metastable state". It is so named because at the given settings in the parameter space both CO-covered and O-covered states are stable and undergo transitions from one to the other seemingly at random. In other words, each state is stable if covering the whole surface, but unstable to fronts between adjacent CO and O covered regions. One example for this behaviour is taken from Lauterbach et al.

and reproduced in fig. 3. It shows a slow progressing CO-front moving in from the left side. The front has on its foremost edge a very bright fringe just a few μm wide, presumably due to the diffusion of CO into the O covered region.

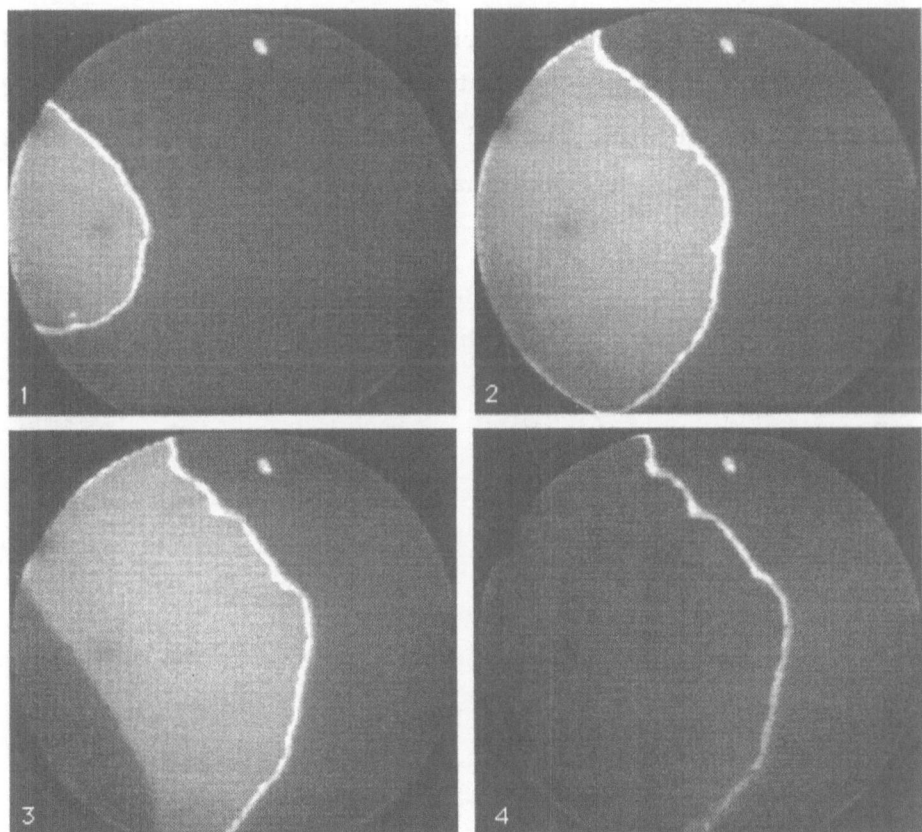

Figure 3. Oxygen covered areas dark, CO covered ones light grey and nearly clean fringes bright; diameter of the pictures: 600 μm, T = 475 K, $P_{CO} = 1.8x10^{-5}$ mbar; $P_{O_2} = 4x10^{-4}$ mbar, relative times 0 s, 15 s, 20 s, 22 s.

The reaction between these species and the desorption of CO_2 leaves a nearly clean area behind before it is covered by CO again. The front speed is approximately 8 μm/s. In the third frame a much faster moving O-front appears and overtakes the slower CO-front. The O-front shows no particular fringe and will annihilate the CO-front within the next second (not shown). At this stage the surface is predominantly O covered and the maximum of productivity , i.e. the highest CO_2 output is reached. The O-front has a speed of 110 µm/s, more than 10 fold faster than the CO-fronts. This causes the corresponding oscillation rates for CO_2 production to be very irregular and they are normally non-symmetric showing a steep increase for CO_2 compared with a less steep decrease. This big difference in front speeds can be explained by a large variance of 3 orders of magnitudes for the sticking probability of O_2 between the reconstructed surface and the bulk-like 1x1 surface. This might also explain why other patterns like spirals or regular standing waves have not been found.

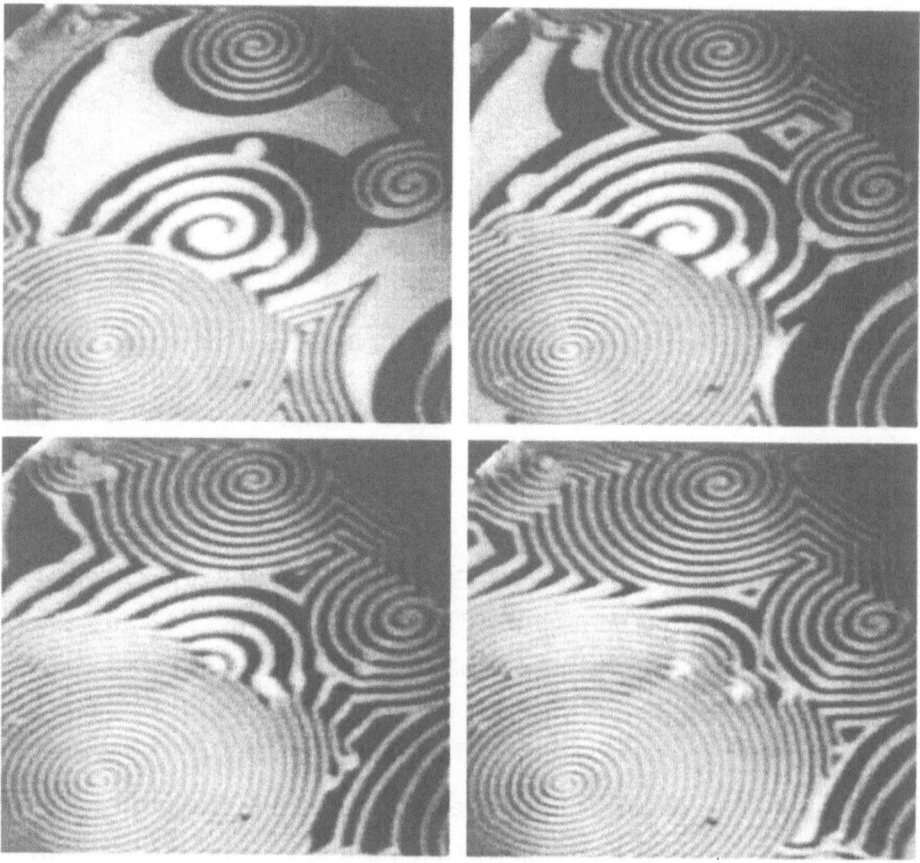

Figure 4.   *Frame size 440 x 410 µm², T = 448 K, P_{CO} = 4.3 x 10⁻⁵ mbar;*
P_{O_2} = 4 x 10⁻⁴ mbar, time difference between the frames 30 s.

On the Pt(110) surface, the difference for the sticking probability of O_2 on the two phases are only 50% compared to about 3 orders of magnitude difference for Pt(100). Therefore the double metastable state occurs in a more limited region of the parameter space and the patterns look more regular. An overview about the patterns found for the (110) surface is given by Nettesheim et al..

From this work a nice example for spirals pinned and defined by different defect sizes is reproduced in fig. 4. The fig. 4 illustrates a population of spirals with greatly different rotation periods and wavelengths. The elongated shape of these spirals is caused by the anisotropic CO diffusivity. The core sizes vary between 25 x 14 µm for the central spiral in the first frame to only 5 x 3 µm for the fastest one in the left lower corner. This spiral, rotating every 6.8 s is about 5 times faster than the central one, annihilating all other spirals and finally dominates the whole image.

Observations like this are most common, but with a slight modulation of one parameter free moving spirals have also be found. A sequence is shown in fig. 5, where the temperature has been varied by about 1 K with a frequency of $w = 0.05 \ s^{-1}$.

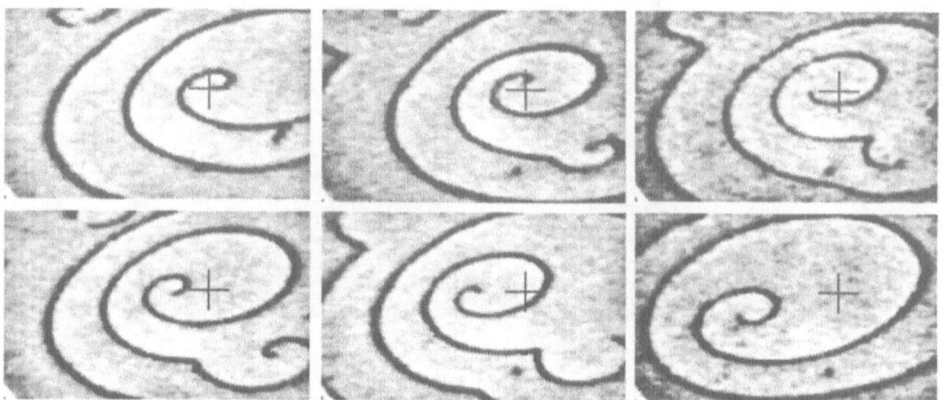

Figure 5.   Frame size 256 x 156 μm^2 , $T = 398$ K, $P_{CO} = 1.8$ x 10^{-5} mbar; $P_{O_2} = 4$ x 10^{-4} mbar, time difference between each frame: 40 s.

The cross in the frames represents the central position of the core in the initial frame for reference. The drift of the core is clearly seen in the sequence. Since the modulation frequency was chosen to nearly 1/2 of the rotation frequency for the spiral, a fairly high drift velocity of 0.35 μm / s compared to the spiral front speed of 0.9 μm / s was found.

In addition to the drifting of the spiral, the influence of defects can again be seen. In the second frame of fig. 5 the spiral arm of the central spiral has broken up presumably by a defect, and a second spiral has been formed, interacting with the old one. The interaction with different wave fronts created by various defects on the surface complicates the picture.

To reduce the complexity and to study the influence of boundary conditions we created by microlithography areas of defined geometric properties consisting of

Figure 6.   Frame size $720 \times 500 \ \mu m^2$, $T = 440 \ K$, $P_{CO} = 5.0 \times 10^{-5}$ mbar; $P_{O_2} = 4 \times 10^{-4}$ mbar

catalytic active Pt surrounded by inactive Ti

A large domain pattern in a three letter-shaped region on Pt(110) is reproduced in fig. 6. The letters are mainly CO covered, only the "H" shows spiral waves. Due to uneven illumination the "I" appears darker than the "F", although both are CO covered. The spiral at the upper corner of the "H" sends its wave fronts downwards where they curve around into the horizontal part of the "H" only to move upwards again in the right part.

Clearly the influence of the higher diffusion rate for CO along the [1,-1,0] direction compared to the [001] direction of the Pt(110) surface is again apparent in the elliptical shapes of those spiral waves.

To study the influence of the anisotropic CO diffusion on Pt(110) further, a more intricate design was chosen (fig. 7). Two perpendicular sets of 20 stripes of active areas were constructed, each one 20 µm wide and separated by a line of 5 µm of Ti. They were approximately aligned along the direction of faster diffusion for CO and orthogonal to it. A scratch running through the upper half of the stripes serves as a nucleation center and the CO waves start from it. Even though the steady state under the chosen condition is a completely CO covered one, the direction of those reaction - diffusion fronts is inversed several times (as indicated by the arrows), apparently in a undecided fashion. The CO fronts sometimes even return all the way back to the scratch where they originated. When they have grown long enough to reach the opposite boundary of the channel, they never again retreat and finally the predominantly CO-covered state prevails. The stripes in the lower half of the frames need much longer to be completely covered by CO, a fact which might be due in part to the missing nucleation center (scratch) and the slower diffusion coefficient for CO along the [0,0,1] direction.

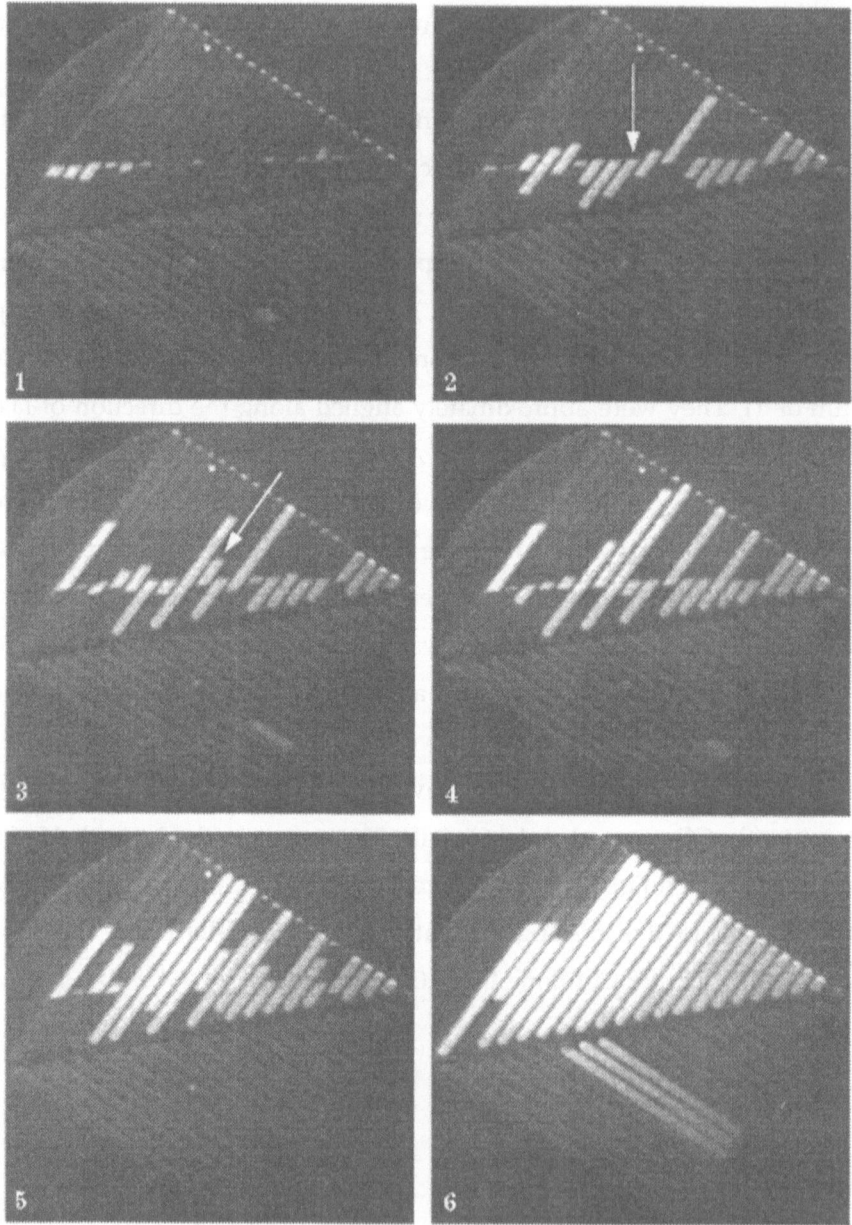

Figure 7.   Time difference between frames 120 s, T = 440 K,
$P_{CO} = 2.3 \times 10^{-5}$ mbar; $P_{O_2} = 4 \times 10^{-4}$ mbar,

To illustrate the role of anisotropy we created a set of concentric rings and started a single CO pulse in one of those rings. Its movement around this ring is depicted in fig. 8. The dimensions of the active Pt(110) ring are inner ø = 25 µm outer ø = 40 µm. Nearly half of one cycle is shown, the full anti-clockwise rotation took 37 s. The pulse being roughly quadratic with 8 x 8 µm² in frame 1, becomes elongated due to the higher diffusion along the [1,-1,0] axis which is oriented from the lower left corner towards the upper right one.

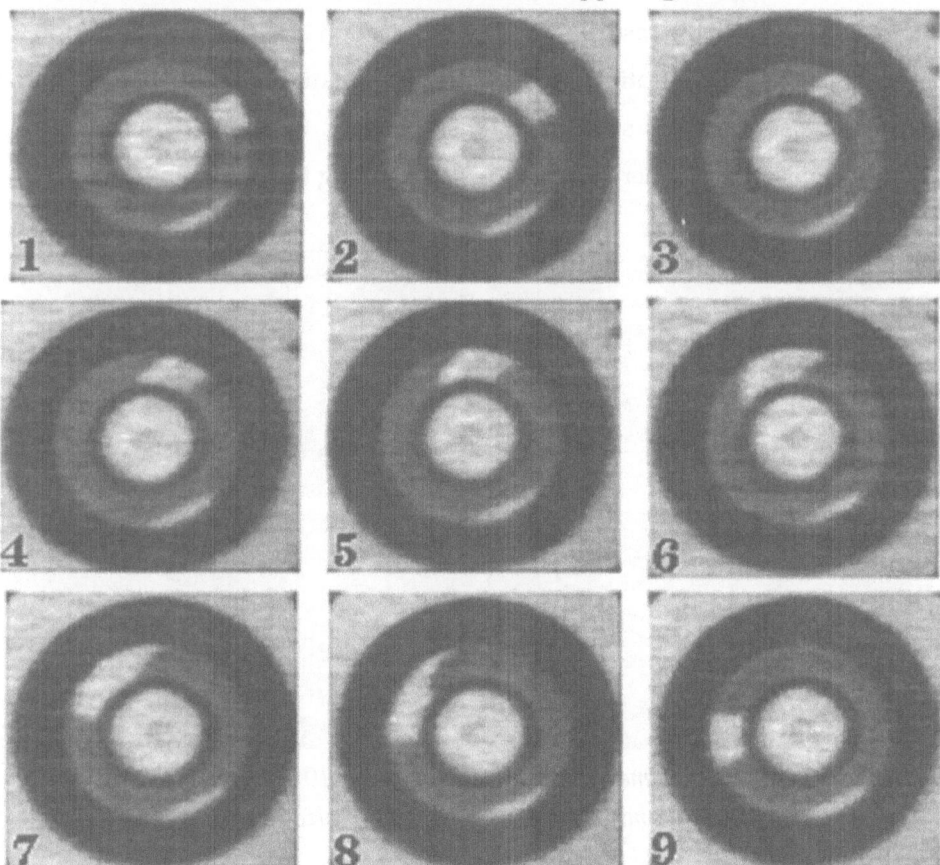

Figure 8.   Time difference between frames 2 s, T = 469 K,
$$P_{CO} = 5.6 \ x \ 10^{-5} \ mbar; \ P_{O_2} = 4 \ x \ 10^{-4} \ mbar,$$

The length of the pulse more than doubles, only to shrink again when it changes its route towards the slower axis in frame 9. A small defect can be seen in the lower right part of the Pt ring, but it does not interfere with the performance of this slightly out of tune "clock". The pulse rotated for many minutes and was surprisingly stable despite increasing or decreasing of the CO partial pressure of up to 5 %. In simulations, diffusion anisotropy is often suppressed by a simple rescaling of the coordinates in such a way that the diffusion appears isotropic. This is of course only possible where no special geometry has been chosen. In the case of the ring, the simulation was performed using a 2-dimensional diffusion operator, resulting in very good agreement [21]. An even more complicated behaviour for the surface diffusion has been observed in the case of the NO + H2 reaction catalysed by Rh(110) where instead of the elliptical spiral patterns, rectangular shaped ones had been found. The authors simulated those patterns by introducing a state-dependent anisotropy, where a variation occurs along the concentration profile of the chemical wave.

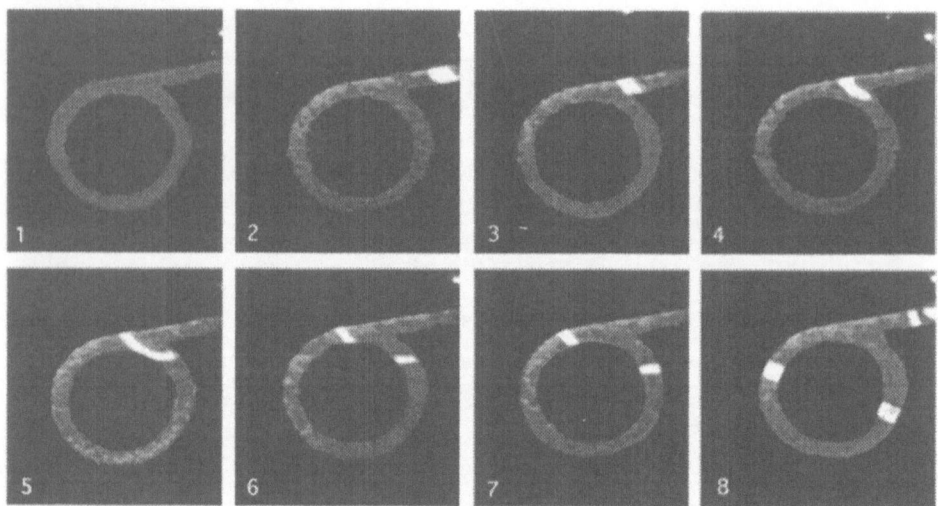

Figure 9. *Time difference between frames 10 s, T = 470 K,*
$P_{CO} = 5.6 \times 10^{-5}$ mbar; $P_{O_2} = 4 \times 10^{-4}$ mbar, size 150 x 150 μm^2

Another example using inert boundary arrangements, but in a slightly more complex geometry than that of fig. 8 is demonstrated in fig. 9.

This synchrotron-like structure was used to study the collisions of solitary waves. Through the injection channel CO pulses travel towards the junction with the ring, where it splits into two parts which then collide head on at the lower part of the ring structure. The intention was to confirm instances of soliton-like behaviour, which had been reported earlier and modelled a short time later. The model was not only capable of simulating the experiments, but also predicted various kinds of wave-splittings which were later found on the video tapes of the earlier experiments. Within the synchrotron-like structures, all collisions studied so far show either the annihilation of one of the pulses or the "explosion" of them starting at the collision point. In the latter case the system changed into the CO covered state (not shown in fig. 9). A different approach uses active boundary conditions like Pd surrounding the single crystalline Pt areas. This inverts the no-flux condition for Ti as surrounding material into a flux condition from the Pd, revealing a new phenomena. in fig. 10. Previous observations of pulses in narrow channels have displayed square shapes, but in this case they are exhibiting a circular like shape presumably due to the flux of CO from the Pd covered areas surrounding the Pt channel.

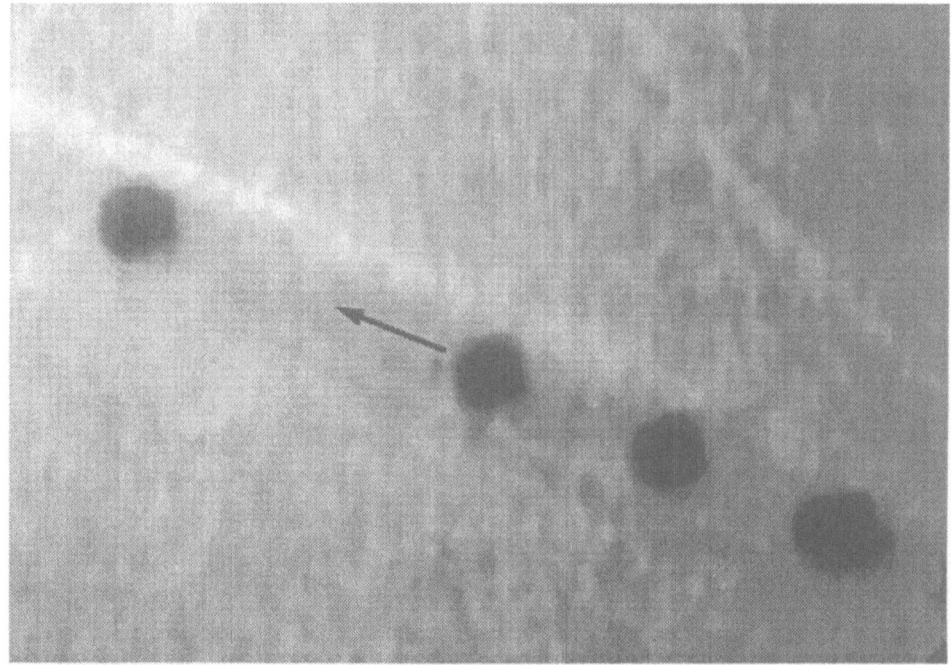

Figure 10.   *Round oxygen pulse travelling in a 20 μm wide Pt(110) channel,*
T = 425 K, P_{CO} = 2.8 x 10^{-5} mbar; P_{O_2} = 4 x 10^{-4} mbar,

The rest of this section will be devoted to the newly developed imaging techniques, EMSI and RAM. As mentioned earlier both are capable of visualising concentration profiles of adsorbates on surfaces with sub-monolayer sensitivity over pressures ranging from UHV to the reality of catalysis at atmospheric or even higher pressures. To explore the capabilities of EMSI we designed it in such a way that the PEEM could operated simultaneously. Therefore we had to use a mirror and a lens inside the UHV to establish the about 70 degrees incident angle needed, adjusting the system to achieve a optimised EMSI picture.

In fig. 11 simultaneously recorded images are reproduced. Although the scales and the orientation for both pictures are different and the EMSI picture is slightly distorted due to the oblique light path, the main features of the CO wave fronts are clearly visible.

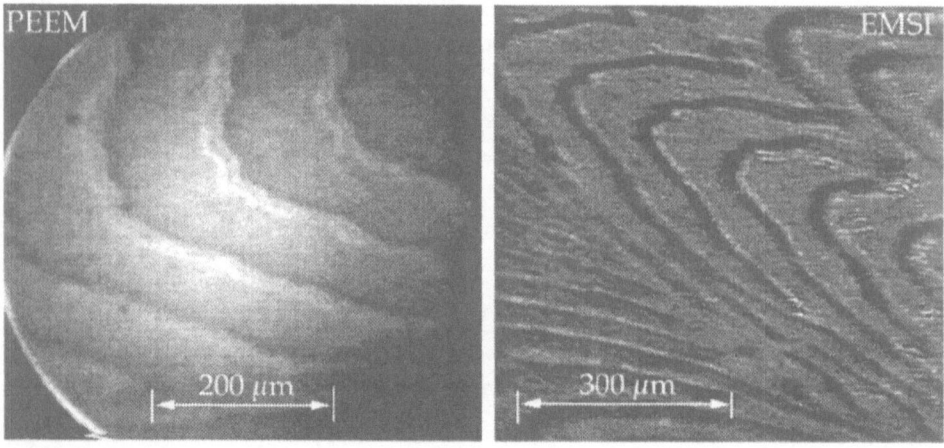

Figure 11.   PEEM and EMSI simultaneously recorded, T = 530 K,
$P_{CO} = 1.4 \ x \ 10^{-5} \ mbar; P_{O_2} = 4 \ x \ 10^{-4} \ mbar,$

In PEEM pictures the CO with its lower work function always shows up lighter than O areas, while with EMSI as well as with RAM the contrast is arbitrary. Depending on the settings of the compensator and analyser, extinction will be for the CO or O covered surface, and therefor the reaction fronts will be darker or lighter. For example in fig. 11 extinction was adjusted for a CO covered surface. Consequently after lowering the CO partial pressure to the appropriate value, the CO-waves are now dark and the greyish areas in between represent mainly O covered regions. The backsides of the CO-fronts display a bright fringe in the EMSI image indicating a different O coverage for that area, an effect also visible for the PEEM picture but not as pronounced.

At pressures much higher than had been possible before we found new features in pattern formation. Fig. 12. represents one example of these findings where we chose a total pressure about hundred times higher than appropriate for PEEM.

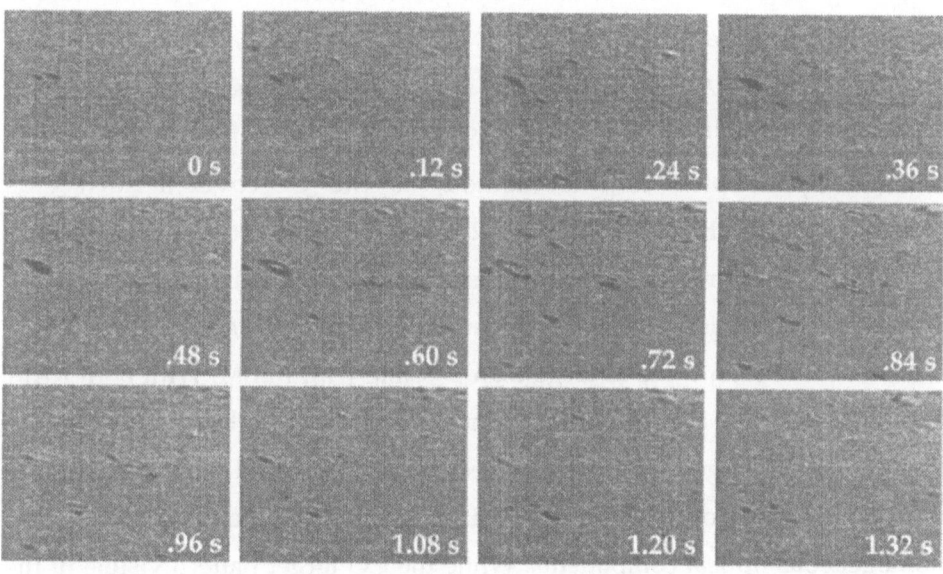

Figure 12.   Series of EMSI images displaying raindrop-like patterns. T = 534K
The diagonal is 1.4 mm., P_{O_2} = 0.022 mbar, P_{CO} = 0.007 mbar.

Watching these images on video strongly reminds the viewer of raindrops falling into a quiet lake. Just a couple of rings appear, then the damping takes over. The "raindrops" appear at random. We believe this remarkable behaviour is due to superposition of reaction-diffusion and thermo-kinetic effects.

At lower pressures target patterns periodically emanate from fixed trigger centers (presumably surface defects) and propagate continuously [28]. The pronounced damping of wave propagation as reported here, and the role of non-isothermal effects in this system will have to be analysed in the future by detailed theoretical modelling. The intensity of this "rain" undulates like in a real rain shower and can be fairly stable for many minutes. After a while, probably due to slight fluctuations of the parameters, the raindrops suddenly expand until the whole surface is covered by CO. This is accompanied by a rapid drop in sample temperature of about 6 K. The speed of cooling is enhanced by convection which already sets in at pressures above 10^{-3} mbar. Since the total pressure at the end of the reaction is slightly higher than at the beginning, cooling by convection is also greater, therefore the temperature is nearly 1 K lower than at the start.

Fig. 13 displays a RAM image recorded at typical conditions for PEEM imaging i.e. 4×10^{-4} mbar of O_2 , The CO oxidation on a Pt(110) sample seems particularly adapted to this method, as the CO-covered surface is unreconstructed while it exhibits a 1x2 missing row structure when oxygen-covered. A difference in anisotropy is thus expected between CO- and O-covered regions. It is noteworthy to mention the size of the frames, 3×4 mm^2 , which is more than twice as large than possible with PEEM. With both RAM and EMSI, the total crystal area including boundaries may be observed in full. Identifying the large target patterns seen in fig. 13 would have been a quite difficult task previously. Note also that in the initial stage we could not use homogeneous optical components. While the CO target pattern visible in the upper right half appear black they are imaged as a bright pattern in the lower left part of the figure. Here the null was adjusted for the homogeneously O-covered surface; however this could not be achieved for the complete image, but only along a diagonal region from the upper left to the lower right corner. Obviously the used optical components were neither free of polarisation effects, nor did they change the polarisation homogeneously across the full image. Since in addition to nulling the intensity we also subtracted an averaged background image from each video frame (that of the O covered surface) the diagonal shows no structures while the same target pattern switches from dark to bright across the diagonal.

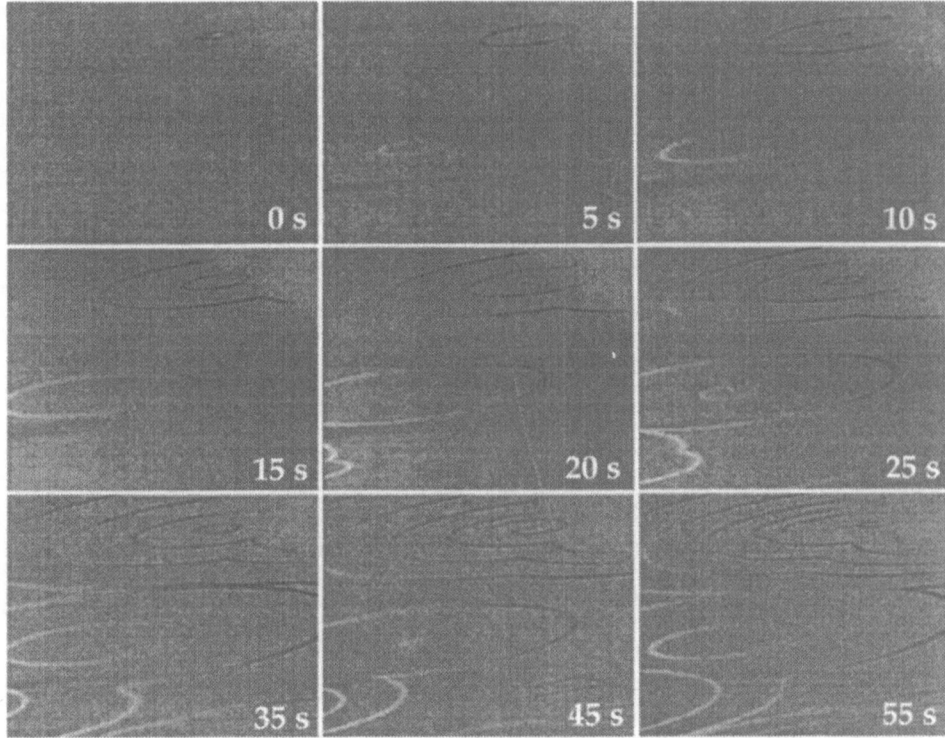

Figure 13.   RAM images recorded at similar conditions as fig. 11, T = 494 K, $P_{CO} = 6.2 \times 10^{-5}$ mbar; $P_{O_2} = 4 \times 10^{-4}$ mbar, image sizes are 3.1 x 3.9 mm^2,

As already supposed in [14] the application of EMSI and RAM imaging the same area of a sample simultaneously might provide supplementary information. This has been proved by Haas et al., where it was shown that the phase transition of the surface structure lags behind the reaction front. This time difference from several to fractions of seconds is of course dependent of the temperature.

Conclusions:

Surface reactions even though they happen on a microscopic scale often exhibit mesoscopic to macroscopic spatio-temporal concentration patterns on the surface. During the last 5 years we have developed several different experimental methods capable of imaging concentration patterns during surface reactions. Here I have highlighted various aspects of pattern formation on the Pt catalysed CO-oxidation. The PEEM is a very versatile technique to image distributions of work function differences and it has potential to improve resolution even to nm scale. On the other hand it has been shown, that the new methods of EMSI and RAM are well suited to image surfaces with a sensitivity to monolayers. A major advantage is their applicability under high pressure conditions. While EMSI might be applied to most systems, RAM can only be applied in samples which exhibit an anisotropy in the optical properties, either by surface reconstruction or by an anisotropic ordered superstructure of the adsorbate and/or their combination.

The pattern formation during catalytic CO oxidation at a Pt(110) sample was investigated up to pressures of 100 mbar revealing distinct differences in the isothermal and non-isothermal regime. At the transition between the two regimes raindrop-like patterns, completely novel in heterogeneous catalysis, have been observed.

Acknowledgements:

The author is indebted to R. M. Tromp, G. Haas, R. U. Franz, J. Lauterbach, A.v. Oertzen, S. Nettesheim, M. Bär, K. Asakura, M. D. Graham I. G. Kevrekidis and G. Ertl for fruitful collaboration in these studies.

The author also acknowledges B.E. Argyle and W. Schrittenlacher for numerous useful suggestions and illuminating discussions.

References :

[1]   Hugo, P. (1970) Stabilität und Zeitverhalten von Durchfluß-Kreislauf-Reak- toren, Berichte der Bunsengesellschaft für physikalishce Chemie 74, 121

[2]   Ertl, G., Norton, P. R. and Rüstig, J. (1982) Kinetic oscillations in the platinum-
        catalysed oxidation of CO, Phys. Rev. Lett. 49, 177-180

[3]   Eiswirth, M., Schwankner, R. and Ertl, G. (1985) Conditions for the occurrence   of kinetic oscillations in the catalytic oxidation of CO on a Pt(100) surface,   Zeitschrift für Physikalische Chemie 144, 59-67

[4]   Eiswirth, M. and Ertl, G. (1986) Kinetic oscillations in the catalytic CO
        oxidation on a Pt(110) surface, Surf. Sci. 177, 90-100

[5]   Cox, M. P., Ertl, G. and Imbihl, R. (1985) Spatial self-organization of surface
        structure during an oscillating catalytic reaction, Phys. Rev. Lett. 54, 1725-1728

[6]   Rotermund, H. H., Ertl, G. and Sesselmann, W. (1989) Scanning photoemission microscopy of surfaces, Surface Sci., 217, L383-L390

[7]   Rotermund, H. H., Jakubith, S., Oertzen, A. v. and Ertl, G. (1989) Imaging of   spatial pattern formation in an oscillatory surface reaction by scanning photo-   emission microscopy, Journal of Chemical Physics 91, 4942-4948

[8]   Rotermund, H. H., Engel, W., Kordesch, M. and Ertl, G. (1990) Imaging of   spatio-temporal pattern evolution during carbon monoxide oxidation on Pt,   Nature 343, 355-357

[9]   Engel, W., Kordesch, M. E., Rotermund, H. H., Kubala, S. and Oertzen, A. v. (1991) A UHV-compatible photoelectron emission microscope for applications   in surface science, Ultramicroscopy 36, 148-153

[10]   Scholz, S. M., Mertens, F., Jacobi, K., Imbihl, R. and Richter, W. in press,
        Reflectance anisotropy on a metal surface: Rhodium(110), Surf. Sci.

[11]   Carroll, J. J. and Melmed, A. J. (1969) Ellipsometry-LEED study of the adsorption of oxygen on (011) tungsten, Surf. Sci. 16, 251-264

[12]   Carroll, J. J., Madey, T. E., Melmed, A. J. and Sandstrom, D. R. (1980) The room temperature adsorption of oxygen, hydrogen and carbon monoxide on   ruthenium: An ellipsometry-LEED characterisation, Surf. Sci. 96, 508-528

[13]   Hall, A. C. (1969) A century of ellipsometry, Surf. Sci. 16, 1-13

[14]   Rotermund, H. H., Haas, G., Franz, R. U., Tromp, R. M. and Ertl, G. in press, Imaging pattern formation in surface reactions from ultra-high vacuum to   atmospheric pressures, Science

[15]   Rotermund, H. H., Haas, G., Franz, R. U., Tromp, R. M. and Ertl, G. in press, Imaging Pattern Formation: Bridging the Pressure Gap,   Applied Physics A

[16]   Rotermund, H. H., Engel, W., Jakubith, S., Oertzen, A. v. and Ertl, G. (1991), Methods and application of UV photoelectron microscopy in heterogeneous catalysis, Ultramicroscopy 36, 164-172

[17]   Aspnes, D. E., Harbinson, J. P., Studna, A. A. and Florez, L. T. (1987) Optical-   reflectance and electron-diffraction studies of molecular-beam-epitaxy growth   transients on GaAs(001), Phys. Rev. Lett. 59, 1687-1690

[18]   Gritsch, T., Coulman, D., Behm, R. J. and Ertl, G. (1989) Mechanism of the   CO induced 1x2 - 1x1 structural transformation of Pt(110), Phys. Rev. Lett. 63

        1086-1089

[19]   Lauterbach, J. and Rotermund, H. H. (1994) Spatio-temporal pattern formation   during the catalytic oxidation of CO on Pt(100), Surf. Sci. 311, 231-246

[20]   Nettesheim, S., Oertzen, A. v., Rotermund, H. H. and Ertl, G. (1993) Reaction   diffusion pattern in the catalytic CO-oxidation on Pt(110); front propagation and   spiral waves, Journal of Chemical Physics 98, 9977-

9985

[21] Graham, M. D., Kevrekidis, I. G., Asakura, K., Lauterbach, J., Krischer, K., Rotermund, H. H. and Ertl, G. (1994) Effects of Boundaries on Pattern Formation: Catalytic Oxidation of CO on Platinum, Science 264, 80-82

[22] Graham, M. D., Bär, M., Kevrekidis, I. G., Asakura, K., Lauterbach, J., Rotermund, H. H. and Ertl, G. (1995) Catalysis on microstructured Surfaces: Pattern Formation during CO oxidation in complex Pt domains, Phys. Rev. E 52, 76-93

[23] Oertzen, A. v., Rotermund, H. H. and Nettesheim, S. (1994) Diffusion of carbon monoxide and oxygen on Pt(110): experiments performed with the PEEM, Surf. Sci. 311, 322 - 330

[24] Gottschalk, N., Mertens, F., Eiswirth, M. and Imbihl, R. (1994) Chemical waves in media with state-dependent anisotropy, Phys. Rev. Lett. 73, 3483-3486

[25] Mertens, F. and Imbihl, R. (1994) Square chemical waves in the catalytic reaction NO + H2 on Rh(110) surface, Nature 370, 124-126

[26] Rotermund, H. H., Jakubith, S., Oertzen, A. v. and Ertl, G. (1991) Solitons in a Surface Reaction, Phys. Rev. Lett. 66, 3083 - 3086

[27] Bär, M., Eiswirth, M., Rotermund, H. H. and Ertl, G. (1992) Solitary-wave phenomena in an excitable surface reaction, Phys. Rev. Lett. 69, 945 - 948

[28] Jakubith, S., Rotermund, H. H., Engel, W., Oertzen, A. v. and Ertl, G. (1990) Spatio-temporal Concentration Patterns in a Surface Reaction: Propagating and Standing Waves, Rotating Spirals, and Turbulence, Phys. Rev. Lett. 65, 3013- 3016

[29] Haas, G., Franz, R. U., Rotermund, H. H., Tromp, R. M. and Ertl, G. in press, Imaging Surface Reactions With Light, Surf. Sci.

Subject Index